Spatial patterns in landscape archaeology

A GIS procedure to study settlement organization in early Roman colonial territories

Spatial patterns in landscape archaeology

A GIS procedure to study settlement organization in early Roman colonial territories

Anita Casarotto

Leiden University Press

This book is the result of a project conducted by A. Casarotto at the Faculty of Archaeology of Leiden University. The project was part of the umbrella project 'Landscapes of Early Roman Colonization' (LERC), funded by the Netherlands Organization for Scientific Research (NWO, project number 360-61-040) and coordinated by Dr. T.D. Stek (Faculty of Archaeology, Leiden University) and Dr. J. Pelgrom (Royal Netherlands Institute in Rome – KNIR). NWO, Leiden University, and KNIR supported the research and production of the present book.

LERC

Landscapes
of Early Roman
Colonization

NWO Nederlandse Organisatie
voor Wetenschappelijk Onderzoek

Universiteit
Leiden

KNIR KONINKLIJK
NEDERLANDS
INSTITUUT
ROME

Archaeological Studies Leiden University is published by Leiden University Press, The Netherlands
Series editors: M.E.R.G.N. Jansen and H. Kamermans

Cover Image: The landscape of Venosa (Basilicata, Italy) with the Monte Vulture in the background
Cover design: J.F. Porck
Lay out: J.F. Porck
Illustrations: A. Casarotto
Image editor: J.F. Porck

ISBN 978 90 8728 311 7
e-ISBN 978 94 0060 328 8
NUR 682

Printed and bound by CPI Group (UK) Ltd, Croydon, CR0 4YY

This book is distributed in North America by the University of Chicago Press (www.press.uchicago.edu).
Printed and bound by CPI Group (UK) Ltd, Croydon, CR0 4YY

Contents

Acknowledgments

First and foremost, I would like to express sincere gratitude to my supervisors, Dr. Tesse D. Stek and Dr. Jeremia Pelgrom. They have been the main source of inspiration for my research. Tesse and Jeremia patiently guided me towards construction of a solid methodology for this research. Their insightful inputs, constructive feedback, and our fruitful discussions have been vital for this research and for my intellectual development. Thanks to Tesse and Jeremia, I feel that my knowledge and archaeological skills have reached a level that I could have never accomplished alone. I cannot thank them enough for their advices and help. Their continuous encouragement and support have been for me like powerful injections of motivation throughout these years, and thus the fundamental factors for completing this book.

Tesse D. Stek and Jeremia Pelgrom are also the directors of the research project within which this work has been developed: the Landscapes of Early Roman Colonization project (LERC) which has been carried out thanks to a Free Competition grant provided by the Netherlands Organization for Scientific Research (NWO, project number: 360-61-040). I would like to thank the Faculty of Archaeology of Leiden University for having been the ideal base where I could conduct my research. Besides being an extraordinary academic organization, the Faculty of Archaeology has been for me like home during these years. I owe sincere gratitude to my LERC colleagues and friends for the support and substantial help: Dr. Marleen Termeer (especially for the extremely useful feedback on the introduction, conclusions, third chapter, and for the Dutch translation of the summary), Rogier Kalkers, Dr. Jesús García Sánchez, Arthur Hamel, and Dr. Gianluca Cantoro.

I would like to thank the Royal Netherlands Institute in Rome (KNIR) for the generous support and the opportunity to profit from the research facilities available at the Institute and in Rome. My research has considerably benefitted from the stimulating and fascinating atmosphere that the KNIR (for me a magic and unique place to work) offers to students and scholars.

I also wish to thank the Leiden University Fund/Mr. J.J. van Walsem Fund and NWO (LERC project) for sponsoring a research trip to the Joukowsky Institute for Archaeology at Brown University, USA.

Another person who had a key role for my intellectual development is Dr. Hans Kamermans. I will never forget his warm hospitality when I joined the Faculty of Archaeology in Leiden. I would like to immensely thank him for his support and help in several matters. Hans instilled me with the desire to constantly improve my skills in computer-based modeling techniques, and he advised me on GIS, predictive modelling and land evaluation analysis.

The time I spent surveying with the LERC team in Venosa and the surrounding municipalities of Basilicata (Italy), has been one of the greatest archaeological fieldwork experiences I have ever had. I enjoyed the good company of a wonderful survey team of international students and academics who worked really hard to collect data that have been fundamental for this project. During this experience, I have learned so much from the survey expertise of Jeremia Pelgrom, and I will always remember our explorative (and quite adventurous!) car trips searching for the right survey area. They were really happy times. Also, the survey expertise of Tesse D. Stek, Jesús García Sánchez, and Arthur Hamel has been important for the successful realization of the survey campaigns in Basilicata. The LERC survey campaigns in Basilicata and Molise were financially supported by NWO, Leiden University, and the KNIR, and were made possible by the *Soprintendenze* of Molise and Basilicata, as well as various local municipalities and institutions in both Molise and Basilicata, and together permitted me to achieve several goals set out by this project, as well as to test preliminary hypotheses literally on the ground.

My sincere gratitude goes also to Professor Jan Sevink and Ruud H. van Otterloo who taught me key principles of soil science, which proved to be fundamental in my research. Their training both in the field and during discussions in Leiden has sparked in me a strong passion for geoarchaeology and soil science.

I am grateful to Professor Peter Akkermans, the members of the doctorate committee, and all the anonymous referees who reviewed my papers for their valuable and constructive comments. I also want to express my gratitude to Jessica Palmer for revising the English writing of a major part of this book.

I would like to warmly thank the Graduate School of Archaeology for the support (especially the director Prof. Miguel John Versluys and the coordinator Roswitha Manning), and my colleagues (and friends) at Leiden University for the stimulating discussions on archaeological topic, and, of course, for the good fellowship especially during lunch at the Faculty of Archaeology. In particular, Dr. Stephanie Döpper, Mark Locicero, Sufyan Al Karaimeh, Roberto Arciero, and Dr. Richard Veymiers. A special thanks goes to my friends Dr. Chiara Piccoli and Dr. Hanna Stöger for the joyful moments we shared, the support and for being helpful in many matters during my stay in Leiden (also outside of office hours).

Further acknowledgments to persons and institutions that supported my work are placed in each chapter.

Outside my professional context, I had the pleasure to meet special persons during my stay in Leiden, with whom I shared really happy times. There is one person in particular who I thank from the bottom of my heart, and this is my dear friend Shan. Since my arrival in The Netherlands, Shan has brought so much joy to me: her friendship is the greatest gift that has been given to me while living in The Netherlands. Without her support, completing this journey would certainly have been impossible. Speaking about meeting special persons, I want to also thank my friends Frida Pellegrino, Giacomo Fontana, Giulia, Teddy, Emanuele, Gertrudis, Mario, and Nathalie for the support, help, and all the joyful moments we shared.

Despite the distance that set us apart, I have always been encouraged by relatives and friends from Italy and spread across Europe. In particular, I would like to thank Alessandra, Angelica, Marcella, Stefania, Laura, Chiara, Viviana, Serena, Davide, Sara, Mattia, Aunt Maria Rosa, Uncle Mario, Uncle Attilio and Grandmothers Maria.

Finally, I want to thank my mom, my dad, and my brother for their unconditional support during this lengthy process. I am so lucky and proud to have them in my life. Last in this list but first in my heart, I thank my beloved Stefano for the encouragement and the exceptional support, for loving me and for being the beautiful person he is.

Chapter 1. Introduction

Material remains of ancient settlements in and on the soil provide important empirical evidence of activities carried out in the past by the ancient communities that settled and transformed the landscape. Through field survey, archaeologists can detect a surface sample of what once existed. However, these archaeological traces are exposed to the elements and constantly run the risk of being destroyed or displaced. Also their visibility may be changed by post-depositional processes (Schiffer 1987; Leonardi 1992; De Guio 1995). The number, density and pattern of sites as documented by surveyors in the field is therefore inherently biased, and not directly representative of ancient human behavior. Before the collected survey data can be used to reconstruct ancient settlement patterns, thorough checks and balances need to be made. The aim of this book is to provide an approach to these checks and balances by developing a set of interrelated GIS analyses that form a research procedure for the use of legacy survey data in settlement pattern analysis (Parsons 1972; Balkansky 2008; Drennan 2008).

The GIS procedure presented in this book is devised to be widely applicable, and thus may be useful to several landscape archaeology projects that deal with (legacy) site-based survey datasets. However, to describe the steps which make up the method, and to substantiate the theory, a case-study is used. Specifically, this book focuses on a recent landscape archaeology project that investigates patterns in early Roman colonial territories of Italy (a description of this project is offered in section 1.5). The available legacy survey data and the research question posed by this project (described in section 1.5) form a key factor for the development and application of the methodological procedure proposed in this book.

Landscape archaeologists concur that GIS is the ideal platform for the analysis of settlement patterns inferable from survey data (Kvamme 1989; Allen *et al.* 1990; Lock & Stančič 1995; Gillings *et al.* 1999; Wheatley & Gillings 2002; Kay & Witcher 2005; Conolly & Lake 2006; Kantner 2008; but see Sharon *et al.* 2004). GIS facilitates the identification of correlations and associations between the environment and the spatial distribution of archaeological data by combining them in an overlay format. This enables significant data patterns to be identified, which helps providing a better understanding of the archaeological problem under study. Since its emergence in archaeology in the early 1990s, GIS has facilitated the integration and comparison of large survey datasets on an inter-regional level, by allowing for integrated digitalization, dynamic spatial visualization, and efficient comparative testing of data. This book shows the crucial role of GIS in the development of methods that use (legacy) survey data in a proper way, and also shows how these methods are key for the future of settlement pattern archaeology. GIS-based procedures will undoubtedly continue to be developed further and gain importance in archaeological pattern analysis, especially as the huge volume of available data resulting from recent fieldwork will surely trigger more GIS experimentation.

1.1 SURVEY ARCHAEOLOGY AND REGIONAL ANALYSIS: A FOCUS ON ITALY

From the 1950s onwards, a widespread exploitation of land for agriculture across the Mediterranean has brought a large amount of archaeological material to the surface (see discussion in Ammerman 1981;

Barker & Llyod 1991; Barker & Mattingly 1999 – 2000; Barker 1996; Bintliff & Sbonias 1999; Pasquinucci & Trement 1999; Piccarreta & Ceraudo 2000; Alcock & Cherry 2004a; Witcher 2008; Attema *et al.* 2010, 1 – 26; Campana 2018, 1-26). Over the last decades, data have been collected in several regions by systematically line-walking (mainly) arable fields. Italy is a prime example of a Mediterranean country where survey has been widely applied to collect archaeological data (see *e.g.* Ward-Perkins 1961; Potter 1979; Yntema 1993; Barker 1995; Marchi & Sabbatini 1996; De Guio *et al.* 1999; Carandini *et al.* 2002; Stek *et al.* 2015; Traviglia 2016; Vermeulen *et al.* 2017; for other Mediterranean countries see Wilkinson 2000; Cherry 2003; Alcock & Cherry 2004a). This has resulted in impressively large datasets that can be used to address a whole range of different archaeological and historical issues. Despite the persistent debate on the validity of field survey as a prospection technique, it is still the most widely applied approach to the study of regional landscapes (Alcock & Cherry 2004a; Galaty 2005; Kantner 2005; Kowaleski 2008; Peterson & Drennan 2011; Feinman 2015).

Conventionally, surface concentrations of artifacts with recognizable limits or sufficiently high sherd density are translated by surveyors into units of analysis, that is archaeological sites (for a discussion on site definition see Plog *et al.* 1978, 365; Gallant 1986; Cherry *et al.* 1991, 45 – 47; Alcock & Rempel 2006). The recorded archaeological sites are subsequently described in databases and represented as points or polygons on topographic maps. The inventories of archaeological sites compiled during field surveys (*i.e.* site-based survey datasets) have an important role in heritage preservation strategies. In Italy for example, the *Forma Italiae* topographic survey program has been producing large registers of sites and monuments that can be visualized in comprehensive archaeological distribution maps (Castagnoli 1978; Azzena & Tascio 1996; Guaitoli 1999; Quilici & Quilici Gigli 2004, 64-80; Sommella 2009; Marchi 2016). These maps can be consulted for several regions across the Italian peninsula and provide valuable tools to modern landscape planners in their decision-making process (for a summary list of survey projects in Italy see Cambi & Terrenato 1994, 21 – 30; Pelgrom 2012; Corazza *et al.* 2014).

In addition to the cultural resource management function, site-based survey datasets have been used to address historical questions. The availability of legacy survey data from different regions encourages integration and cross-comparison of settlement patterns on an inter-regional level, which is uniquely valuable for research focusing on large-scale cultural processes (*e.g.* Patterson *et al.* 2000; 2004; Attema & van Leusen 2004; Patterson 2004; Thompson 2004; Palmisano *et al.* 2018; for Mediterranean examples outside Italy, *e.g.* Alcock 1993; Stone 1997; Blanton 2000; Wilkinson 2000; 2003; Diacopoulos 2004; Bintliff *et al.* 2007; Farinetti 2011). With regard to Italy, and Italian Roman contexts in particular, these data have been used to study ancient settlement patterns (*e.g.* Pelgrom 2008; Witcher 2008; Kay & Witcher 2009; Stek *et al.* 2015; 2016, 36 – 38), demographic trends, and economic strategies of past societies (*e.g.* Goodchild 2007; 2013; Witcher 2005; Wilson 2008; Fentress 2009; Launaro 2011; Pelgrom 2012; 2013) in a diachronic perspective.

The GIS research procedure proposed in this book is devised to use site-based survey datasets for research purposes (although it can also be useful for cultural resource management). Specifically, it can be used to implement a sound regional-scale investigation of settlement patterns based on a site dataset. Additionally, this method can be useful for the comparative analysis of site datasets from different regions. To show its applicability for both regional pattern analysis and inter-regional comparative pattern analysis, a case-study is presented in this book which is based on legacy site-based survey datasets of regions in central-southern Italy. This analysis aims to improve the understanding of the settlement developments triggered by an important large-scale cultural phenomenon that affected the Italian peninsula during the 3rd - 1st century B.C., namely Roman conquest and colonization.

1.2 THE IMPORTANCE OF LEGACY SURVEY DATA

Legacy survey data can be defined as the archaeological information registered in the early decades of archaeological survey (1950s – 2000s), when survey methodology was in development and had not yet reached the methodological sophistication of today's surveys and techniques (Witcher 2008). Usually these legacy datasets consist of registers of archaeological sites which are only roughly described and represented as symbols (*i.e.* points or polygons) on topographic maps. Devising a method for realizing the potential of these legacy survey data is extremely important, especially as recent studies have convincingly argued that the archaeological surface record is disappearing very rapidly (Knudson 1989; Cameron 1994; Wilkinson 2000; Cherry 2003, 157; Kowalewski 2008, 231 – 232 with further references; Renfrew & Bahn 2008, 559-570; Witcher 2008; Campana 2018). Modern anthropogenic transformations (triggered by the intensification of farming practices and urban expansion) and related geomorphological processes (erosion, deposition) are acutely affecting the soil and thus pose major threats to the preservation of the archaeological record contained on and under the surface (Boardman & Poesen 2006; Witcher 2008). It is likely that legacy data will soon be the only available source of information to study past settlement dynamics for many regions.

Survey has been traditionally conceived as a repeatable experiment (but see discussion in Ammerman 1981; Lloyd & Barker 1981). Ideally, archaeologists could come back into the field if necessary and walk it again to collect more or better data. Especially for Italy, this scenario has turned out to no longer be possible in several areas. The increased deforestation and urbanization, and the intensive plowing and land leveling for agriculture, endanger the soil layer by causing extensive soil loss in many farmed and urban areas of the world (see discussion in Borselli *et al.* 2006; Torri *et al.* 2006; Van Oost & Govers 2006; WWF 2017). The extent to which this landscape transformation is taking place is remarkable, and for Italy previously unparalleled (see discussion in Torri *et al.* 2006; ISPRA 2017). This situation, of course, also has a detrimental effect on the survival, identification, quality, and reliability of the archaeological material recordable nowadays at the surface level. In many places, material scatters have disappeared or have been dramatically damaged. During our recent survey campaigns in southern Italy (in 2013 – 2016, see Chapter 5) we had the opportunity to assess directly in the field how dramatic this situation is, and we concluded that a reduced quality of the material collectable at the surface by today's surveys is to be expected. As this trend in agriculture is unlikely to stop in the near future, and its changes to the landscape geomorphology are irreversible, archaeologists will need to depend more and more heavily on surveys that have been conducted in the past. For this reason, there is a growing interest in formal procedures to use the available legacy data systematically.

In addition to the threat on the surface record, interest in systematic procedures has also been stimulated by the rapid and widespread sharing of digitalized legacy survey data. Thanks to open access web repositories for data dissemination[1], survey data are becoming more and more accessible. This increased availability of digital datasets was already noted by Alcock and Cherry in their seminal introduction to the book *Side-by-side survey* (2004a). They also stressed the fact that "[...] we are faced with an influx of regionally based archaeological data of unprecedented quantity, quality, and diversity from hundreds of individual survey projects – but no widely agreed procedures for juxtaposing, combining, or synthesizing individual survey datasets" (Alcock & Cherry 2004b, 4). This book helps filling in this gap by providing a GIS-based procedure for the systematic and comparative analysis of existing regional survey datasets to tackle specific large-scale research questions.

1 *E.g.*: MAGIS project (http://cgma.depauw.edu/MAGIS/), ARIADNE project (http://www.ariadne-infrastructure.eu/), DANS project (https://dans.knaw.nl/nl), MOD project (http://mappaproject.arch.unipi.it/mod/Index.php), Fasti Online Survey project (http://www.fastionline.org/survey/). At Leiden University, the Leiden Centre of Data Science (LCDS, https://www.universiteitleiden.nl/en/science/leiden-centre-of-data-science) is a network of researchers from different disciplines (*e.g.* computer science, statistics, mathematics, law, natural sciences, and also archaeology), who explore methods to manage the large amounts of data that are becoming available (Big Data) also in archaeology (see discussion in Gattiglia 2015).

1.3 THE INTEGRATION OF LEGACY SURVEY DATASETS FOR INTER-REGIONAL COMPARATIVE ANALYSIS

There are two main problems when using legacy survey data for inter-regional analysis of settlement patterns. One problem, which affects both regional and inter-regional investigations, regards the fact that these data are often not accurate enough to perform advanced modeling and formal statistical analysis (see discussion in Crema 2015). The second issue concerns the many methodological differences between individual surveys and the results they produce. The variegated nature of legacy survey datasets challenges integration and thus has significant limitations for inter-regional comparison of settlement patterns (see discussion in Mattingly & Witcher 2004; Witcher 2008; Launaro 2011, 85-88; Fracchia 2013, 185 – 190; Witcher & Craven 2013).

The need is pressing for the development of methods that enable comparative analysis of existing and new datasets (Cherry 1983; Kintigh 2006; Witcher 2008; Kintigh *et al.* 2014; Feinman 2015). However, despite the recent intense theoretical discussions within the archaeological community on the unexploited potential of legacy data on the one hand, and their many weaknesses on the other, there is a lack of practical methodological solutions which can accommodate the innate problems and eventually enable correct integration for inter-regional comparisons (but see discussion and possible solutions in Van Leusen 2002; Attema & van Leusen 2004; Allison 2008; Witcher 2008; Aloia *et al.* 2017).

A significant obstacle for the comparison of these datasets is the lack of sufficient metadata, that is data describing data properties regarding, for instance, site definition criteria or survey strategy adopted by survey teams to collect these data (Wise & Miller 1997; Witcher 2008). Many of these datasets were collected using different methods that were often only generically described. This lack of detail in methodological information limits the possibilities to compare different regional datasets. Additionally, the data quality varies a good deal between surveys. These qualitative and methodological differences between surveys are extremely difficult to overcome and significantly challenge systematic integration and comparison of data properties across regions (Ammerman 1981; Witcher 2008; Attema *et al.* 2010, 19 -29). Hence, it is important to underline that a fundamental condition for the correct application of the procedure presented in this book lies in the comparability of the site-based survey datasets selected for the comparison (see Chapter 3).

Thus far only few solid research procedures to use these datasets appropriately have been devised (*e.g.* Van Leusen 2002; Attema *et al.* 2010; Feiken 2014)[2]. One of the reasons for this is the lack of comparable regional datasets, namely datasets that are comparable because they were collected using the same (or sufficiently similar) survey methodologies and described with a common terminology. To overcome this lack of compatibility, scholars are investigating solutions to make more datasets comparable. Several researchers are currently building formal ontologies and standard vocabularies that will allow integration of a higher number of (legacy) survey datasets. These specialists make a serious effort to try to arrive at a shared consensus on standardization procedures for survey methods and survey datasets[3]. If successful, such ambitious projects will greatly impact inter-regional comparative pattern analysis, and will surely trigger the development of new procedures to formally use survey datasets in settlement pattern

2 Scholars are well-aware that the production of concrete methods should be undertaken now. This is testified by the high number of recent conferences and interdisciplinary collaborations that aim to stimulate discussion on methods for data preservation and (re)use. Just to name an example, the most important and well-known international conference on computer applications and quantitative methods in archaeology (CAA) activated in 2012 the CAA Recycle, a yearly award to encourage scholars using legacy data to share fruitful ideas about the best-practice on data reuse methods and technologies (http://caa-international.org/bursaries/recycle-award/).

3 An interesting project aiming at integration of legacy and new datasets of the suburbium of Rome using standard ontology, was presented in a session organized by Peter Attema, Paolo Carafa, Willem Jongman, and Christopher Smith at the RAC conference in Rome, March 17th 2016 (session 1). Another interesting roundtable session on Conceptual Reference Models (CIDOC CRM approach) was organized by Martijn van Leusen, Tymon de Haas, and Achille Felicetti at the EAA conference in Maastricht, September 1st 2017 (roundtable session 251). Tymon de Haas and Martijn van Leusen (De Haas & Van Leusen 2016) also directed a pilot project funded by DANS (Data Archiving and Networked Services) and the Groningen Institute of Archaeology (GIA, University of Groningen) that aimed to use the CRM approach for integrating archaeological survey datasets.

investigations. The current research shows that when comparable survey datasets are available it is possible to successfully apply the GIS procedure described below (section 1.4) and to shed light on ancient settlement patterns both on a regional and on an inter-regional level. Thus, this book aims to provide a solid research procedure for the use of comparable (legacy) survey datasets (section 1.4).

1.4 OUTLINE OF THE PROPOSED RESEARCH PROCEDURE

1.4.1 PREMISES

The proposed research methodology is based on two fundamental premises. Firstly, (legacy) survey data are intrinsically biased and incomplete, and thus a formal research procedure is required to cope with the limitations and fragmentary nature of this particular type of data. Another fundamental problem is that legacy survey data are usually not collected and described with the detail that sophisticated modeling techniques require to perform sound statistical analysis. A procedure must be developed that can deal with these uncertainties and shortcomings in the source datasets (see also discussion in Orton 2004, esp. section 4.2). Most importantly, the research procedure cannot be technologically-driven, using for instance readymade GIS packages and sophisticated statistical models borrowed from other disciplines, but should, instead, be data-driven. If the data type does not allow for the application of sophisticated statistical tools, we should not use these tools. It is always tempting to apply innovative quantitative tools for the analysis of survey data. However, as these tools are often developed to analyze datasets of very different quality and nature, there is a real risk that such advanced modeling techniques lead to misguided results when it comes to archaeological survey data.

Therefore, it is crucial to develop an analysis strategy that is based on the survey data and their particular properties. As a rule, this requires the development of tailor-made GIS-based strategies, in which only a selection of suitable quantitative and qualitative methods can be applied. This process requires a lot of preliminary experimentation before the best suited methodology surfaces for answering specific research questions[4]. Cutting-edge modeling techniques, when performed solely by virtue of their level of technological or statistical innovation, do not necessarily guarantee that a research question can be tackled correctly (see discussion in Stek 2016). As a preliminary step, therefore, a careful examination of both data quality and research goals is needed, followed by the selection of suitable methods that do justice to the potential of the data as well as addressing their problems.

The second premise regards the issue of comparability between surveys and the data they produce. In this book legacy site datasets have been selected that were collected using comparable recording methods (*e.g.* survey methodology, space interval between surveyors) and site criteria (*e.g.* sherd threshold for site detection; scatter size; dating method). In this way, the effect of methodological differences in survey strategies is limited and can be assumed (at least theoretically) not to be responsible for possible inter-regional variations in the spatial patterns inferable from the site distributions of different regional datasets. Still, there are certain site criteria that are subjectively defined by survey teams and, as a rule, are uncertain and vary considerably from survey to survey, and thus from region to region. For example, functional classifications of site types (*e.g.* farm, hut, villa, hamlet, village) are notoriously difficult to establish and metadata describing type definition criteria convincingly does not as yet exist. Differences in definitions used between surveys are thus very common. These differences are extremely difficult to resolve and constitute an obstacle for data

4 Experiments with inductive predictive modeling were carried out at the beginning of this project. This technique predicts zones of expected high and low settlement suitability based on the correlation between known sites and landscapes (Kohler & Parker 1986; Judge & Sebastian 1988; Kvamme 1990; Kamermans & Wansleeben 1999; Verhagen 2007). Since this technique highlights attractive and repulsive ecological units for settlement within the landscape, I initially used it as a means to investigate ancient settlement strategy (for this approach in another Roman context see *e.g.* Goodchild 2007, 121 - 178). However, because of the nature of the available source datasets, I realized that in doing so I was providing hypothetical reconstructions that worked only under the assumption of representative samples and environmental determinism in ancient location preferences. To put forward sounder reconstructions a testing of these assumptions was in order. Therefore, I decided to use a different strategy and to develop a procedure that incorporates such a testing.

integration and comparison on the level of functional site types. Therefore, in this book such qualitative functional categories are disregarded. Instead, only the more formally established site properties such as site position, size, and chronological range of occupation have been included in the various performed GIS analyses.

1.4.2 PROCEDURE

As will become clear in the following chapters, it is possible to produce useful results through GIS strategies that are based on the above described premises, especially in the case of problem-oriented regional investigations. The best results are attained when using a research methodology that consists of two parts. One part regards the assessment of the degree of distortion characterizing our data (*i.e.* the application of bias-testing approaches). The other part aims to extract historical meaning from these data. This part constitutes the problem-solving phase of the procedure: through the application of different approaches, it aims to shed light on the archaeological question that is posed by a project. It is important to stress that the order in which these two operations (the bias-testing and problem-solving part) are carried out is flexible and depends very much on the case-study and on the question being asked.

The problem-solving part of the procedure consists of three subsequent, but interrelated analytical steps. The first step is a deductive analysis in which various existing theoretical models on settlement patterns and location preferences are used as testable hypotheses. These models are confronted with a survey dataset in order to establish which one is the most reliable for that particular territory (see Chapter 3). As a second step, a bottom-up, that is inductive, analysis is performed to establish if unexpected patterns emerge, which cannot be explained by the initial working hypothesis (Chapter 6). The analysis starts from the data and highlights patterns in site distributions with respect to the local environment. As a third step, the results of the deductive analysis are compared to the results of the inductive analysis[5]. It is at this stage that new insights can be gained about the initial theoretical settlement models, namely on how these models functioned in specific regions (see Chapter 7).

There is a long history of discussion in archaeological theory on which type of scientific reasoning is the most appropriate to derive knowledge: deductive or inductive (Salmon 1976; Kamermans & Wansleeben 1999; Wheatley 2004; Verhagen 2007; Kamermans *et al.* 2009; Bintliff & Pearce 2011; Verhagen & Whitley 2012; Citter 2012; Casarotto 2015). This opposition between deductive and inductive reasoning, however, is only a construct; both approaches are equally valid, and should be combined to achieve a solid interpretation (Glaser & Strauss 1967; Kamermans 2000; Löwenborg 2010; Verhagen & Whitley 2012, 50 – 55; De Guio 2015, 301; Arnoldus *et al.* 2016). This book, indeed, is a further demonstration that such approaches do not oppose one another, but on the contrary, are complementary. Through the combined application of deductive and inductive methods, and the consequent dialectic argumentation on the results, it was possible to acquire new information from the initial theory and the initial data. What follows now is a more detailed explanation of each of the three steps composing the problem-solving part of the research procedure.

The first step is defined as a theory-driven approach (deductive analysis). It consists of the testing of existing theories on ancient settlement patterns using legacy survey data. In this phase, these data are used simply as a touchstone for testing the tenability of theories. In order to do so, the settlement theories must first be properly described and translated into archaeological surface scenarios. This can be achieved through an analytical exercise. To do this, the archaeological evidence (detectable at the surface by means of field survey) associated with the theoretical models under consideration must be established. In other words, the theoretical models and the site patterns projected by them (*i.e.* the expected empirical output), are retraced to the

5 Besides its utility for settlement pattern analysis, this deductive-inductive procedure has also proved to be useful for archaeological predictive modeling (Verhagen & Whitley 2012; Arnoldous *et al.* 2016; Patacchini & Nicatore 2016).

archaeological correlates upon which they are based. Then, those correlates that are retrievable from the available legacy survey data, and therefore are testable, can be selected. These correlates play a crucial role as potential indicators for the existence of certain models of settlement organization in antiquity. Especially those empirical correlates that, according to competing settlement theories, would manifest themselves differently, or in total opposition, in the archaeological surface record play a key role in verifying the models' reliability (*e.g.* regular or clustered site distribution; low or high site density). By systematically analyzing the selected empirical correlate using suitable GIS quantitative and qualitative tools we can determine which model(s) most likely occurred in the past (for more details on this indicator-based approach see Stek 2016, 151 - 158).

As a second step, the data are tested for patterns using an inductive approach. Whereas the deductive approach is mainly descriptive with respect to patterns exhibited by site-based data, the inductive analysis is, instead, exploratory. The goal in this case is to explore whether relationships can be found between site-based survey data and a set of local landscape characteristics (*i.e.* environmental and cultural factors characterizing the local context). This eventually enables the gaining of precise insights into the local settlement system behind settlement patterns of a particular region. Significant local correlations spark further understanding on the settlement rationale underlying survey data, and thus enrich the initial theory on ancient settlement systems. By indicating possible variants in settlement location strategy, this inductive analysis prompts researchers to calibrate or refine existing general starting models (namely, those tested in the previous deductive analysis), and to assess how they functioned with respect to the specific historical and ecological region under consideration (see below).

Such a calibration constitutes the third phase of the problem-solving part of the procedure. The calibration occurs in at least three different ways. When there is accordance between the results from the deductive and inductive analyses, the existing settlement theory acquires more credibility. If, instead, discordance exists, partial or total reconsideration of previous theories is in order. In the case in which the inductive analysis discloses patterns that are not in opposition with the general patterns expected by a previous theory, but that need further explanation, researchers may be prompted to fine-tune such a theory in more detailed terms, and for the specific context under consideration. Moreover, the comparison between the results from the deductive and inductive analyses potentially enables moving from observations of patterns to contextual interpretations of these patterns (*i.e.* aspects of past settlement strategies in specific regions) (Kantner 2005, 1182). Through a process of validation and refinement of an existing theory, this procedure may eventually advance previous knowledge about ancient settlement strategies (see Chapters 6 and 7).

From the above, it is argued that this deductive-inductive research method allows the gaining of new knowledge from a previously existing theory. When the initial theory is disproved because unexpected patterns are found in the data, this triggers the development of new theories. The same goes for the opposite case. If the initial theory has been proven to be reliable, but is still underdeveloped since there is insufficient insight into certain aspects, this approach can enrich it, possibly filling some gaps with supplementary correlations inferred from the data (Verhagen & Whitley 2012, 90). Eventually, these additions to the extant knowledge lead to new theory, and the deductive-inductive cycle can be activated again: the theory fine-tuned with newly acquired insights about certain aspects can be tested, in other regions as well, and if proven to be reliable and leading to new correlations, the researcher may try to feed it back again, with other significant observations inferred from the data (this is seen as the cycle of research, Tashakkori & Teddlie 2003).

However, the deductive-inductive research method described above must be supplemented with bias testing. Survey data are intrinsically biased, and to propose an accurate territorial reconstruction it is necessary to test whether these data are representative. That is to say, it is necessary to test whether (legacy) survey data, from which ancient settlement patterns can be assessed both deductively and inductively, are affected by distorting factors such as surface visibility and geomorphological processes of erosion and deposition. The research procedure proposed in this book incorporates methods for the evaluation of biases that may alter the regional configuration of archaeological sites in a landscape. The position of this bias testing within the procedure is flexible; we will see that for the specific case-study analyzed in this book, it was performed after the deductive and before the inductive analysis in order to follow the logical flow of the argumentation[6] (Chapters 4 and 5). However, this is not a rule, and researchers may decide, according to the questions they are tackling, when it is the proper moment to carry out the test.

The full integration within the research procedure of GIS methods to assess visibility and geomorphological biases involved in regional field surveys, is the greatest strength of the proposed methodology. Through these integrated methods archaeologists can assess, while performing regional analysis, the degree of incompleteness of their datasets and thus further calibrate their interpretations about settlement patterns. In this way, sound historical reconstructions can hopefully be put forward.

1.5 CASE STUDY

To show how the proposed research procedure works in practice, a particular case-study is considered in this book. This case-study regards a landscape archaeology project that aimed to test two different settlement theories by using (mainly) survey data. These two competing theories describe the impact of early Roman colonization on rural settlement patterns in central and southern Italy in a very different way.

The LERC archaeological project (Landscapes of Early Roman Colonization, led by Dr. Tesse D. Stek and Dr. Jeremia Pelgrom), which this PhD research is part of[7], aimed to test a new conception of Roman Republican colonization. This project targeted non-urban settlements, such as villages and hill forts, to gain understanding of their societal role during the early Roman expansion in Italy (Pelgrom 2008; Stek & Pelgrom 2013; Pelgrom *et al.* 2014; Stek *et al.* 2015; Stek *et al.* 2016). Traditionally, the settlement organization of colonial landscapes is imagined to have been characterized by the presence of small colonists' farms regularly and densely distributed across the countryside (in a centuriated area) around a colonial town (*e.g.* Brown 1980; Rathbone 1981; 2008; Celuzza & Regoli 1982; Settis 1984). The existence of such an agriculture-based city state model has long been conventionally accepted in discourses about Roman colonization (Salmon 1969; see discussion in Pelgrom & Stek 2014; Pelgrom 2014; Stek 2014). LERC challenges this view by testing survey data for the existence of nucleated colonial settlement systems (*i.e.* villages) in several territories affected by Roman colonization prior to the Second Punic War (before 218 – 201 B.C.). In this context, the PhD research aimed at assessing the viability of these two competing settlement theories. The research question to be tackled using legacy site-based survey data was: which is the most probable settlement model for the organization of colonial rural landscapes?

With this research goal in mind, the GIS-based procedure described in section 1.4 was developed and applied to this case study. As a first step, a testable archaeological correlate for which the two settlement theories expected different spatial patterns was selected (see also Stek 2016). To be suitable for rigorous comparative testing, this archaeological correlate had to have been measured by the previous surveys using comparable methods and criteria (*i.e.* similar survey methodology, sherd threshold for site detection, space interval between surveyors, and dating method). Based on these parameters, point distributions representing settlement sites

6 The theories tested in the deductive analyses were based on the same biased legacy datasets; in the context of testing their reliability it was therefore crucial to initially use the datasets as they stand, without any bias filtering.

7 I conducted the PhD project at the Faculty of Archaeology of Leiden University from 2013 to 2018 thanks to the LERC funds deriving from a NWO free competition grant (Netherlands Organization for Scientific Research, LERC project number 360-61-040, PI Dr. Tesse D. Stek). For further information on the LERC project: https://landscapesofearlyromancolonization.com/
https://www.universiteitleiden.nl/en/research/research-projects/archaeology/landscapes-of-early-roman-colonization

were chosen as a suitable correlate. Specifically, site distributions registered on topographic maps by survey projects carried out in the colonial landscapes of Venusia, Cosa and Aesernia (central-southern Italy) were considered. The main reason these three colonial territories were selected for a comparative analysis of settlement patterns was because the field surveys conducted there were comparable in terms of survey methodology. This way, we could exclude major differences in survey methodologies as the cause for differences in regional patterns between the territories under consideration. Additionally, the selection of sample areas also depended on the fact that these territories historically played a key role in the Roman colonization of the Italian peninsula (3rd – 1st century B.C.) (see Chapter 3).

First, the research question was addressed deductively. That is, the feasibility of existing settlement theories was assessed by noting to what extent survey data, and specifically the selected testable archaeological correlate (*i.e.* point distributions of settlement sites), fitted within the patterns expected from these theories. Specifically, the reliability of the two competing settlement models of Roman colonization was tested against the selected archaeological correlate using GIS-based density and pattern analyses. The first model envisages large colonies of people departing from Rome to occupy wild indigenous territories through the establishment of a fortified urban center, regular land division systems for agricultural production in the surrounding territory, and the construction of numerous farms in the allotted holdings resulting from these partitions. Such an impressive re-organization of the landscape would have, in most cases, destroyed previous settlements or driven off native people into marginal zones where they continued settling in villages (see also Attolini *et al.* 1991, 144 - 145; Marchi & Sabbatini 1996; Carandini *et al.* 2002, 108 – 110). Therefore, scholars supporting this traditional view would expect survey to have detected in these colonial landscapes a high and even distribution of sites, which ought to have been plotted on maps as evenly dispersed points (but see discussion in Chapter 2 and results in Chapter 3).

Due to a lack of archaeological and epigraphic evidence in support of such a scenario, scholars have lately started to question this view, and an alternative settlement model has been proposed. This new model puts forward a revisionist understanding of early colonial landscapes, rejects the conventional view of radical changes in settlement strategies from pre-colonial to colonial periods, and proposes, instead, the presence of a dominant clustered village-based settlement system both in pre-colonial and early Roman colonial periods (4th - 3rd century B.C.) (Pelgrom 2008; Stek 2009; Stek & Pelgrom 2013; Stek 2018). According to this theory, survey in colonial territories would not record regular site distributions, but rather clusters of archaeological sites. The first analysis presented in this book (Chapter 3) shows that there is a remarkable discrepancy between survey data and the conventional theory that assumes high site density and dispersed patterning. Instead, the archaeological site dataset is better compatible with the clustered configuration proposed more recently. In conclusion, according to a deductive analysis of the raw survey data, the nucleated settlement theory is more probable.

As a second step, however, the question is raised whether the observed nucleated patterns in the datasets may be a result of post-depositional biasing factors, rather than of ancient settlement strategies (Chapters 4 and 5). We investigated erosion and depositional geomorphological processes along with surface visibility conditions of the walked fields, and tested whether these factors could have been responsible for the recording of empty or low site density zones in between clusters of archaeological sites. Testing for distorting factors involved in survey procedures (*i.e.* biases) is crucial to tackle the research question addressed in this case-study. For instance, the low site density documented in a majority of colonial landscapes in central-southern Italy has been often explained by a supposed poor surface visibility of colonial sites (Rathbone 1981; 2008; see discussion in Pelgrom 2008; 2012). The bias-testing could eventually shed light on whether the smoothly dispersed and densely-settled farm landscape expected by conventional views was invisible to modern archaeological surveys (thus it could not be recorded) or, conversely, whether it is unlikely to have ever existed in the early colonial period (3rd century B.C.).

Field survey methods have their implicit limitations, and can offer only a fragmentary image of ancient landscapes. However, as shown in Chapters 4 and 5, the settlement data recorded in two field surveys actually display representative evidence to detect significant regional patterns and large-scale trends in settlement location preferences (Haggett *et al.* 1977; Kantner 2005). These patterns and trends can be highly informative of the social use of space in ancient times (see also the discussion in Barker 1991). The reliability of the original survey data was not only confirmed by modeling techniques performed in GIS, but also by new survey campaigns carried out recently in the context of our project (Pelgrom *et al.* 2014; 2016; Stek *et al.* 2015; 2016). The direct testing of the legacy data and models in the field with new surveys lent convincing support to the conclusions drawn for this case-study (see also van Leusen 2002; Patterson 2004; Patterson *et al.* 2004; Kay & Witcher 2005; Attema *et al.* 2010; De Haas 2011).

In order to gain further understanding on the logic behind the detected settlement clustering, as a next step, a bottom-up, inductive analysis was performed on the legacy survey data (Chapter 6). This not only permitted to further test the highlighted patterns (Chapter 3) but also provided the opportunity to assess whether there were other local correlations which could help explain them for one specific historical and ecological context (Chapter 6). The comparison between deductive and inductive results adds a new explanatory perspective to the lively debate on the interpretation of regional settlement patterns in Roman colonial landscapes. By confronting the results from the array of approaches used in the deductive and the inductive analyses we could note whether or not they corroborated and/or complemented each other. It is especially through this feedback between the deductive and the inductive reasoning that further understanding on settlement patterns in colonial landscapes could be gained.

The comparison between the results from the first-deductive (Chapter 3) and the last-inductive analysis (Chapter 6) showed that the impact of Roman colonization on a newly conquered area of Southern Italy may not have been as harsh as previously imagined. It is possible that, as supported by the LERC nucleated model, the establishment of a colonial population coming from Rome did not immediately lead to radical transformations in the organization of previous landscapes. The colonial-period site clustering displayed in archaeological site distribution maps, whether representing villages or other types of settlement agglomerations[8], grew organically and complemented, rather than replaced, the pre-colonial settlement system in the Republican period (3rd – 1st century B.C.).

1.6 STRUCTURE OF THE BOOK

This book consists of a collection of articles. These articles are the output of my PhD project, which is part of the larger Landscapes of Early Roman Colonization project introduced earlier. A majority of the articles has already been published in peer-reviewed journals during the course of the PhD, and some others are awaiting publication. Each chapter contains one article. Articles are not presented in chronological order of publication but instead follow the logical flow of the proposed research procedure. This basically consists of the methodological steps a researcher may follow in order to implement a regional pattern analysis based on an existing survey site dataset(s). All chapters, except the second, tackle one step. Chapter 2, instead, provides background about the *status quaestionis* on settlement theory regarding the organization of colonial landscapes, the case study presented, and the historical debate followed throughout the book.

1.6.1 CHAPTER 2: FAMILIARIZING WITH THE DEBATE

What would the conventional settlement model for Roman colonial landscapes, in theory, have looked like on a site distribution map?

8 From site point distributions it is impossible to assert whether the detected clustering corresponded to villages, other types of nucleated settlements or loose groups of sites (see discussion in Roberts 1996). Only excavation and invasive prospection techniques (see discussion in Verhagen 2013) could potentially unlock the characteristics of these clusters, thus permitting precise evaluations about their nature and chronology (Pelgrom *et al.* 2014; 2016; Stek *et al.* 2016).

It is difficult to picture how a colonized territory may have appeared using a conventional view without a drawing or a map to illustrate it. It is even more difficult, then, to imagine how archaeological sites representing the surface evidence of colonial-period settlements should have been plotted as dots on a topographic map during pedestrian surveys (*i.e.* our testable archaeological correlate). Since dots-on-a-map is the empirical correlate used in the case-study presented in this book, in Chapter 2, a quantitative GIS method is proposed to simulate the dispersed distribution of archaeological sites expected by the conventional settlement theory. The visualization of this hypothetical landscape permits to visually assess the huge difference in site number and pattern between what was actually recorded during a field survey in the area of Venosa in southern Italy and what was expected to be recorded. The chapter ends by opening up the possibility that, rather than visibility impediments for site detection, other reasons may have determined the recorded site arrangement, such as a different settlement strategy from the one expected.

This topic is treated in Chapter 2 that corresponds to the article:

Casarotto, A. 2017. A method for modeling dispersed settlements: visualizing an early Roman colonial landscape as expected by conventional theory. *Archeologia e Calcolatori* 28 (1), 147-163. https://doi.org/10.19282/AC.28.1.2017.09

1.6.2 CHAPTER 3: DEDUCTIVE ANALYSIS

Which theoretical model fits the colonial site densities and patterns best?

The dispersed and the village settlement models expect radically different settlement densities and patterns in point distributions representing archaeological sites. These expectancies can be tested through GIS point density and pattern tools. In this chapter three colonial landscapes of central-southern Italy are systematically compared, to assess with which settlement model the survey data accord best. This theory-laden analysis indicates which scenario is more likely to have actually existed in the past, and clearly points out which one is improbable. In addition to that, this analysis also shows that the dichotomous interpretative framework (dispersed versus clustered settlement patterning) adopted in this deductive analysis to investigate settlement systems in colonial landscapes may be too narrow: cultural and environmental factors of the local context may have led to the establishment of a more flexible model of settlement occupation.

This topic is treated in Chapter 3 that corresponds to the article:

Casarotto, A., Pelgrom, J. & Stek, T.D. 2016. Testing settlement models in the early Roman colonial landscapes of Venusia (291 B.C.), Cosa (273 B.C.) and Aesernia (263 B.C.). *Journal of Field Archaeology* 41(5), 568-586. https://doi.org/10.1080/00934690.2016.1211474

1.6.3 CHAPTERS 4 AND 5: ASSESSING BIASES

Are the patterns and densities of survey sites the result of biasing factors rather than real location preferences?

It is possible that, to a certain extent, the site point distribution may be patterned by geomorphological and surface visibility biases (*e.g.* Cherry *et al.* 1991; Terrenato & Ammerman 1996; Attema 2017). For example, there may exist an association between favorable surface visibility conditions and the detection of localized high site densities (*i.e.* point clusters). Conversely, in a zone where site density is low and erosion or deposition is active or occurred recently, the lack of archaeological sites may be explained by geomorphological processes affecting the preservation of sites rather than by past constraints against settlements. It is crucial to assess the effect of distorting factors on site discovery when searching for reliable patterns of ancient settlement systems. In these chapters, survey data of the colonial landscapes of Aesernia and Venusia are tested for possible regional biases using two different methodological approaches. For the case of Cosa, this testing is not carried out because the available information on survey methods and visibility conditions is insufficient for the types of analyses applied in these chapters.

Chapters 4 and 5 correspond to the articles:

Casarotto, A., Stek, T.D., Pelgrom, J., van Otterloo, R.H. & Sevink, J. 2018. Assessing visibility and geomorphological biases in regional field surveys: The case of Roman Aesernia. *Geoarchaeology, an International Journal* 33 (2), 177 – 192. https://doi.org/10.1002/gea.21627

Casarotto, A., García Sánchez, J., Stek, T. D. & Pelgrom, J. (forthcoming). Surface visibility and legacy survey data: between desktop-based analysis and new fieldwork. Article submitted to *FOLD&R: the Journal of Fasti Online*

1.6.4 CHAPTER 6: INDUCTIVE ANALYSIS AND CALIBRATION

What is the influence of local environmental and cultural factors on early colonial settlement?

In Chapter 6 the location preferences and patterns inferred from the legacy data by means of inductive descriptive and quantitative analyses are described. The clustering displayed by the survey data is compared to the natural and cultural local environment to shed further light on the rationale behind its spatial arrangement. Specifically, we evaluate to what degree the settlement patterning is influenced by local environmental and cultural factors. This is done only for the colonial landscape of Venusia because for the other two colonial territories the quantity and/or quality of the available data and/or metadata were insufficient to allow the type of analyses implemented in this chapter. However, the regional patterns of the landscape of Venusia could trigger further comparative testing in other colonial landscapes. The insights gained from the settlement patterns in Venusia can thus contribute to the debate on the rationale of colonial settlement (cf. section 1.4). After having assessed the effect of biases (Chapter 5) and having established that survey data retain substantial representative evidence for inferring ancient patterns and location preferences, in the final part of the book a historical interpretation of the settlement behavior behind the patterns is provided. This is done by combining the results obtained in the previous deductive investigation (Chapter 3) with the results of the inductive analysis.

Chapter 6 corresponds to the article:

Casarotto, A., Pelgrom, J. & Stek, T.D 2017. A systematic GIS-based analysis of settlement developments in the landscape of Venusia in the Hellenistic-Roman period. *Archaeological and Anthropological Sciences*. https://doi.org/10.1007/s12520-017-0561-y

1.7 REFERENCES

Alcock, S. E., 1993. *Graecia capta: The landscapes of Roman Greece*. Cambridge: Cambridge University Press.

Alcock, S.E. & Cherry, J.F. (eds.) 2004a. *Side-by-side survey: Comparative regional studies in the Mediterranean world.* Oxford: Oxbow Books.

Alcock, S.E. & Cherry, J.F. 2004b. Introduction. In S.E. Alcock & J.F. Cherry (eds.), *Side-by-side survey: Comparative regional studies in the Mediterranean world*, 1-9. Oxford: Oxbow Books.

Alcock, S.E. & Rempel, J.E. 2006. The more unusual dots on the map: 'Special purpose' sites and the texture of landscape. In P.G Bilde & V. Stolba (eds.), *Surveying the Greek chora. The Black Sea region in a comparative perspective*, 27-46. Aarhus: Aarhus University Press

Allen, K.M.S., Green, S.W. & Zubrow, E.B.W. (eds.) 1990. *Interpreting space: GIS and archaeology*. London, New York and Philadelphia: Taylor & Francis.

Allison, P. 2008. Dealing with legacy data: An introduction. *Internet Archaeology* 24. http://dx.doi.org/10.11141/ia.24.8

Aloia, N., Binding, C., Cuy, S., Doerr, M., Fanini, B., Felicetti, A., Fihn, J., Gavrilis, D., Geser, G., Hollander, H., Meghini, C., Niccolucci, F., Nurra, F., Papatheodorou, C., Richards, J., Ronzino, P., Scopigno, R., Theodoridou, M., Tudhope, D., Vlachidis, A. & Wright, H. 2017. Enabling European archaeological research: The ARIADNE E-Infrastructure, *Internet Archaeology* 43. https://doi.org/10.11141/ia.43.11

Ammerman, A.J. 1981. Surveys and archaeological research. *Annual Review of Anthropology* 10, 63-88.

Arnoldus-Huyzendveld, A., Citter, C. & Pizziolo, G. 2016. Predictivity – Postdictivity: A Theoretical framework. In S. Campana, R. Scopigno, G. Carpentiero & M. Cirillo (eds.), *Keep the revolution going. Proceedings of the 43rd Annual Conference on Computer Applications and Quantitative Methods in Archaeology, Siena 30 March – 3 April 2015*, 593-598. Oxford: Archaeopress.

Attema, P.A.J. 2017. Sedimentation as geomorphological bias and indicator of agricultural (un)sustainability in the study of the coastal plains of south and central Italy in antiquity. *Journal of Archaeological Science: Reports* 15, 459 – 469.

Attema, P.A.J., Burgers, G.-J. & van Leusen, P.M. 2010. *Regional pathways to complexity: Settlement and land-use dynamics in early Italy from the Bronze age to the Republican period*. Amsterdam: Amsterdam University Press.

Attema, P. A. J., van Leusen, P. M. 2004. Intra-regional and inter-regional comparison of occupation histories in three Italian regions: The RPC project. In S. E. Alcock & J. F. Cherry (eds.), *Side-by-side survey: Comparative regional studies in the Mediterranean world,* 86 - 100. Oxford: Oxbow.

Attolini, I., Cambi, F., Castagna, M., Celuzza, M., Fentress, E., Perkins, P. & Regoli, E. 1991. Political geography and productive geography between the valleys of the Albegna and the Fiora in northern Etruria. In: G. Barker & J. Lloyd (eds.), 1991. *Roman landscapes: Archaeological survey in the Mediterranean region*, 142-152. London: British School at Rome.

Azzena, G. & Tascio, M. 1996. Il Sistema Informativo Territoriale per la carta archeologica d'Italia. In M.L. Marchi & G. Sabbatini, *Venusia (IGM 187 I NO / I NE)* (Forma Italiae 37), 281 – 297. Firenze: Leo S. Olschki.

Balkansky, A.K. 2008. Settlement pattern analysis. In: D.M. Pearsall (ed.), *Encyclopedia of Archaeology*, 1978-1980. New York: Academic Press.

Barker, G. 1991. Approaches to archaeological survey. In G. Barker & J. Lloyd (eds.), *Roman landscapes: Archaeological survey in the Mediterranean region*, 1 – 9. London: British School at Rome.

Barker, G. 1995. *A Mediterranean valley: Landscape archaeology and Annales history in the Biferno valley*. London and New York: Leicester University press.

Barker, G. 1996. Regional archaeological projects: trends and traditions in Mediterranean Europe. *Archaeological Dialogues* 3(2), 160-175.

Barker, G. & Lloyd, J. (eds.) 1991. *Roman landscapes: Archaeological survey in the Mediterranean region*. London: British School at Rome.

Barker, G. & Mattingly D.J. 1999 - 2000. *The archaeology of Mediterranean landscapes*. Oxford: Oxbow Books.

Bintliff, J. L. & Pearce, M. (eds.) 2011. *The death of archaeological theory?* Oxford and Oakville: Oxbow Books.

Bintliff, J. L. & Sbonias, K. (eds.) 1999. *Reconstructing past population trends in Mediterranean Europe (3000 B.C.–A.D. 1800)*. Oxford: Oxbow Books.

Bintliff, J. L., Howard, P. & Snodgrass, A. 2007. *Testing the hinterland: the work of the Boeotia survey (1989-1991) in the southern approaches to the city of Thespiai*. Cambridge: McDonald Institute for Archaeological Research, University of Cambridge.

Blanton, R.E. 2000. *Hellenistic, Roman and Byzantine settlement patterns of the coast lands of Western Rough Cilicia*. BAR International Series 879. Oxford: Archaeopress

Boardman, J. & Poesen, J. 2006. *Soil erosion in Europe*. West Sussex, England: John Wiley & Sons, Ltd.

Borselli, L., Torri, D., Øygarden, L., De Alba, S., Martinez-Casasnovas, J.A., Bazzoffi, P. & Jakab, G. 2006. Land Levelling. In J. Boardman & J. Poesen (eds.), *Soil Erosion in Europe*, 643-658. West Sussex, England: John Willey & Sons Ltd.

Brown, F.E. 1980. *Cosa: The making of a Roman town*. Ann Arbor: University of Michigan Press.

Cambi, F. & Terrenato, N. 1994. *Introduzione all'archeologia dei paesaggi*. Roma: La Nuova Italia Scientifica.

Cameron, C.M. 1994. The destruction of the past: nonrenewable cultural resources. *Nonrenewable Resources* 3(1), 6-24.

Campana, S. R. L. 2018. *Mapping the archaeological continuum. Filling 'empty' Mediterranean landscapes*. Springer.

Carandini, A., Cambi, F., Celuzza, M. G. & Fentress, E. 2002. *Paesaggi d'Etruria. Valle dell'Albegna, Valle d'Oro, Valle del Chiarone, Valle del Tafone*. Roma: Edizioni di Storia e Letteratura.

Casarotto, A. 2015. Location preference modeling alias predictive modeling for the study of archaeological contexts. In G. Pizziolo & L. Sarti (eds.), *Predicting Prehistory: Predictive models and field research methods for detecting Prehistoric contexts. Proceeding of the international workshop 19-20 September 2013, Grosseto*, 35-48. Firenze: Museo ed Istituto Fiorentino di Preistoria.

Castagnoli, F. 1978. La carta Archeologica d'Italia. *Parola del Passato* 33, 78–80.

Celuzza, M. G. & Regoli, E. 1982. La valle d'Oro nel territorio di Cosa. Ager Cosanus e Ager Veientanus a confronto. *Dialoghi di Archeologia* 4, 31–62.

Cherry, J.F. 1983. Frogs around the pond: perspectives in current archaeological survey projects. In D. Keller & D. Rupp (eds.), *Archaeological survey in the Mediterranean area* (BAR International Series 155), 375-416. Oxford.

Cherry, J.F. 2003. Archaeology beyond the site: regional survey and its future. In J.K. Papadopoulos & R.M. Leventhal (eds.), *Theory and practice in Mediterranean archaeology: Old world and new world perspectives* (UCLA Cotsen Institute of Archaeology Monograph Series), 137-59. Los Angeles.

Cherry, J. F., Davis, J. L. & Mantzourani, E. 1991. *Landscape archaeology as long-term history. Northern Keos in the Cycladic Islands from Earliest settlement until modern times*. Los Angeles: UCLA Institute of Archaeology.

Citter, C. 2012. Modelli predittivi e archeologia postclassica: vecchi strumenti e nuove prospettive. In F. Redi & A. Forgione (eds.), *VI Congresso Nazionale di Archeologia Medievale (L'Aquila 2012)*, 12-15 Firenze: All'Insegna del Giglio.

Conolly, J. & Lake, M. 2006. *Geographical Information Systems in archaeology*. Cambridge: Cambridge University Press.

Corazza, V., Di Renzoni, A. & Ferranti, F. 2014. Tutti a spasso: dal Crati al Po passando per il Tevere, cinquant'anni di camminate per campi. In: *150 di Preistoria e Protostoria in Italia*, 543-549. Firenze: Istituto Italiano di Preistoria e Protostoria.

Crema, E.R. 2015. Time and probabilistic reasoning in settlement analysis. In J.A. Barcelo & I. Bogdanovic (eds.), *Mathematics and Archaeology*, 314-334. Boca Raton, FL: CRC Press.

De Guio, A. 1995. Surface and subsurface: Deep ploughing into complexity. In W. Hensel, S. Tabaczynski & P. Urbanczyk (eds.), *Theory and practice of archaeological research*, II, 329 – 414. Warszawa: Institute of Archaeology and Ethnology, Commiteee of Pre- and Protohistoric Sciences, Polish Academy of Sciences.

De Guio, A. 2015. Nuove linee di ricerca fra archeologia pre-dittiva e post-dittiva. *Archeologia e Calcolatori* 26, 301-313.

De Guio, A., Whitehouse, R.D. & Wilkins, J. 1999. Progetto Alto-Medio Polesine-Basso Veronese: decimo rapporto. *Quaderni di Archeologia del Veneto* 15, 101-116.

De Haas, T.C.A. 2011. *Fields, farms and colonists: Intensive field survey and early Roman colonization in the Pontine region, central Italy*. Volume 1. Groningen: Barkhuis & Groningen University Library.

De Haas, T.C.A. & Van Leusen, P. M. 2016. *Extending CRMarchaeo for archaeological field survey – a pilot* (Internal report 2016). Groningen: Groningen Institute of Archaeology, University of Groningen.

Diacopoulos, L. 2004. Investigating social complexity through regional survey:

"second-generation" analysis of Bronze Age data from the Canadian Palaipaphos survey project, Southwestern Cyprus. *Journal of Mediterranean Archaeology* 17, 59-85.

Drennan, R.D. 2008. Settlement system analysis. In D.M. Pearsall (ed.), *Encyclopedia of Archaeology*, 1980-1982. New York: Academic Press.

Farinetti, E. 2011. *Boeotian Landscapes: A GIS-based study for the reconstruction and interpretation of the archaeological datasets of ancient Boeotia* (BAR international series, S2195). Oxford: Archaeopress.

Feiken, M. 2014. *Dealing with biases: Three Geo-Archaeological approaches to the hidden landscapes of Italy*. Eelde, Netherlands: Barkhuis.

Feinman, G.M. 2015. Settlement and landscape archaeology. *International Encyclopedia of Social & Behavioral Sciences* (Second edition) Volume 21, 654-658. doi: 10.1016/B978-0-08-097086-8.13041-7

Fentress, E. 2009. Peopling the countryside: Roman demography in the Albegna valley and Jerba. In A. Bowman & A. Wilson (eds.), *Quantifying the Roman economy: Methods and problems*, 127-163. Oxford: Oxford University Press.

Fracchia, H. 2013. Survey, Settlement and Land Use in Republican Italy. In J. DeRose Evans (ed.), *A Companion to the Archaeology of the Roman Republic*, 181-197. Blackwell Publishing Ltd.

Galaty, M. L. 2005. European regional studies: A coming of age? *Journal of Archaeological Research* 13(4), 291-336.

Gallant, T.W. 1986. "Background noise" and site definition: a contribution to survey methodology. *Journal of Field Archaeology* 13(4), 403 – 418. doi: 10.2307/530167

Gattiglia, G. 2015. Think big about data: Archaeology and the Big Data challenge. *Archäologische Informationen* 38, 113-124.

Gillings, M., Mattingly, D. & van Dalen, J. (eds.) 1999. *Geographical Information Systems and Landscape Archaeology*. Oxford.

Glaser, B.G. & Strauss, A.L. 1967. *The discovery of grounded theory: Strategies for qualitative research*. New Brunswick and London: Aldine

Goodchild, H. 2007. *Modelling Roman agricultural production in the middle Tiber valley, Central Italy*. PhD dissertation, University of Birmingham: eTheses Repository. http://etheses.bham.ac.uk/175/

Goodchild, H. 2013. GIS models of Roman agricultural production. In: A. Bowman & A. Wilson (eds.), *The Roman agricultural economy: organization, investment, and production*, 55-83. Oxford: Oxford University Press

Guaitoli, M. 1999. Appendice III. Nota sulla metodologia della raccolta, dell'elaborazione e della presentazione dei dati. In P. Tartara, *Torrimpietra (IGM 149 I NO)*, (Forma Italiae 39), 357-365. Firenze: Leo S. Olschki.

Haggett, P., Cliff, A. D. & Frey, A. P. 1977. *Locational analysis in human geography*. London: Arnold.

Judge, W. J. & Sebastian, L. (eds.) 1988. *Quantifying the present and predicting the past: Theory, method and application of archaeological predictive modeling*. Denver: Bureau of Land Management.

Kamermans, H. 2000. Land evaluation as predictive modelling: a deductive approach. In G. Lock (ed.), *Beyond the map: archaeology and spatial technologies*, 124-146. Amsterdam: IOS Press.

Kamermans, H., van Leusen, M. & Verhagen, P. (eds.) 2009. *Archaeological prediction and risk management. Alternatives to current practice*. ASLU 17. Leiden: Leiden University Press.

Kamermans, H. & Wansleeben, M. 1999. Predictive modelling in Dutch archaeology, joining forces. In J.A. Barceló, I. Briz & A. Vila (eds.), *New techniques for old times. CAA98. Computer applications and quantitative methods in archaeology. Proceedings of the 26th Conference, Barcelona, March 1998* (BAR International Series 757), 225-230. Oxford: Archaeopress.

Kantner, J. 2005. Regional analysis in archaeology. In H. D. G. Maschner & C. Chippindale (eds.), *Handbook of archaeological methods* (Volume 1), 1179-1223. Oxford: Altamira Press.

Kantner, J. 2008. The archaeology of regions: from

discrete analytical toolkit to ubiquitous spatial perspective. *Journal of Archaeological Research* 16(1), 37-81.

Kay, S. & Witcher, R.E. 2005. The Tiber Valley project: The role of GIS and databases in field survey data integration and analysis. *Archeologia e Calcolatori* 16, 113-37.

Kay, S. & Witcher, R.E. 2009. Predictive modelling of Roman settlement in the middle Tiber valley. *Archeologia e Calcolatori* 20, 277-290.

Kintigh, K. 2006. The promise and challenge of archaeological data integration. *American Antiquity* 71(3), 567-578. doi:10.2307/40035365

Kintigh, K.W., Altschul, J.H., Beaudry, M.C., Drennan, R.D., Kinzig, A.P., Kohler, T.A., Limp, W.F., Maschner, H.D.G., Michener, W.K., Pauketat, T.R., Peregrine, P., Sabloff, J.A., Wilkinson, T.J., Wright, H.T. & Zeder, M.A. 2014. Grand challenges for archaeology. *American Antiquity* 79(1), 5-24.

Knudson, R. 1989. North America's threatened heritage. *Archaeology* 42 (1), 71 – 73

Kohler, T. A. & Parker, S. C. 1986. Predictive models for archaeological resource location. In M.B. Schiffer (ed.), *Advances in archaeological method and theory* 9, 397-452. New York: Academic Press.

Kowalewski, S. A. 2008. Regional settlement pattern studies. *Journal of Archaeological Research* 16(3), 225-285.

Kvamme, K.L. 1989. Geographic information systems in regional archaeological research and data management. *Archaeological Method and Theory* 1, 139-203.

Kvamme, K. L. 1990. The fundamental principles and practice of predictive archaeological modelling. In A. Voorrips (ed.), *Mathematics and information science in archaeology: A flexible framework*, 257-295. Bonn: Holos.

ISPRA, 2017. *Consumo di suolo, dinamiche territoriali e servizi ecosistemici*. Rapporti 266/2017. ISPRA, Roma.

Launaro, A. 2011. *Peasants and slaves. The rural population of Roman Italy (200 B.C. to A.D. 100)*. Cambridge: Cambridge University Press.

Leonardi, G. 1992. Assunzione e analisi dei dati territoriali in funzione della valutazione della diacronia e delle modalità del popolamento. In M. Bernardi (ed.), *Archeologia del Paesaggio*, 25 - 66. Firenze: All'Insegna del Giglio.

Lloyd, J. A. & Barker, G. 1981. Rural settlement in Roman Molise: problems of archaeological surveys. In G. Barker & R. Hodges (eds.), *Archaeology and Italian Society*, British Archaeological Reports, International Series 102, Oxford, 289 - 304.

Lock, G. & Stančič, Z. (eds.) 1995. *Archaeology and Geographical Information Systems: a European Perspective*. London: CRC Press.

Löwenborg, D. 2010. *Excavating the digital landscape: GIS analyses of social relations in central Sweden in the 1st millennium A.D.* Aun 42. Uppsala.

Marchi, M.L. 2016. Carta Archeologica d'Italia – "Forma Italiae" project: Research method. In *LAC 2014 Proceedings (Rome 2014)*. doi: http://dx.doi.org/10.5463/lac.2014.42

Marchi, M.L. & Sabbatini, G. 1996. *Venusia (IGM 187 I NO / I NE)* (Forma Italiae 37). Firenze: Leo S. Olschki.

Mattingly, D. J. & Witcher, R. E. 2004. Mapping the Roman world: the contribution of field survey data. In S.E. Alcock & J.F. Cherry (eds.), *Side-by-side survey: Comparative regional studies in the Mediterranean world*,173-186. Oxford: Oxbow Books.

Orton, C. 2004. Mathematical modelling, *Internet Archaeology* 15. http://intarch.ac.uk/journal/issue15/orton_toc.html

Palmisano, A., Bevan, A & Shennan S. 2018. Regional demographic trends and settlement patterns in Central Italy: Archaeological Sites and Radiocarbon Dates. *Journal of Open Archaeology Data*, 6 (2). DOI: https://doi.org/10.5334/joad.43

Parsons, J.R. 1972. Archaeological settlement patterns. *Annual Review of Anthropology* 1,127-150.

Pasquinucci, M. & Trément, F. (eds.) 1999, *Non-destructive techniques applied to landscape archaeology*. The archaeology of Mediterranean Landscapes 4. Oxford.

Patacchini, A. & Nicatore, G. 2016. Potential paths

and historical road network between Italy and Egypt: from predictive to postdictive approach. In S. Campana, R. Scopigno, G. Carpentiero & M. Cirillo (eds.), *Keep the revolution going. Proceedings of the 43rd Annual Conference on Computer Applications and Quantitative Methods in Archaeology, Siena 30 March - 3 April 2015*, 669-681. Oxford: Archaeopress.

Patterson, H. (ed.) 2004. *Bridging the Tiber. Approaches to regional archaeology in the middle Tiber valley* (Archaeological Monograph of the British School at Rome 13). British School at Rome: London.

Patterson, H. & Di Gennaro, F., Di Giuseppe, H., Fontana, S., Gaffney, V., Harrison, A., Keay, S.J., Millet, M., Rendeli, M, Roberts, P., Stoddert, S. & Witcher, R. 2000. The Tiber valley project: the Tiber and Rome through two millennia. *Antiquity* 74 (284), 395-403.

Patterson, H., Di Giuseppe, H. & Witcher, R. 2004. Three South Etrurian "Crises": first results of the Tiber valley project. *Papers of the British School at Rome* 72, 1-36.

Pelgrom, J. 2008. Settlement organization and land distribution in Latin colonies before the Second Punic War. In L. De Ligt & S. J. Northwood (eds.), *People, land and politics. Demographic developments and the transformation of Roman Italy 300 B.C.-A.D. 14*, 317–356. Leiden: Brill.

Pelgrom, J. 2012. *Colonial landscapes. Demography, settlement organization and impact of colonies founded by Rome (4th–2nd centuries B.C.)*. PhD Dissertation, Leiden University: Leiden University Repository.

Pelgrom, J. 2013. Population density in mid-Republican Latin colonies: A comparison between text-based population estimates and the results from survey archaeology. *Atlante Tematico di Topografia Antica* 23, 73–84.

Pelgrom, J. 2014. Roman colonization and the city-state model. In T.D. Stek & J. Pelgrom (eds.), *Roman Republican Colonization. New Perspectives from Archaeology and Ancient History*, 73-86. Rome: Palombi Editori.

Pelgrom, J., Lecce, L., García Sánchez, J. & Stek, T.D. 2016. Le indagini olandesi sul pianoro di Masseria Casalini Sottana (2013 - 2014). In A. De Siena & T. Giammatteo (eds.), *Palazzo San Gervasio. Modalità insediative e pratiche funerarie dal territorio*, 277 – 288. Venosa: Osanna Edizioni

Pelgrom, J., Marchi, M.L., Cantoro, G., Casarotto, A., Hamel, A., Lecce, L., García Sánchez, J. & Stek, T.D. 2014. New approaches to the study of village sites in the territory of Venosa in the Classical and Hellenistic periods. *Agri Centuriati* 11, 31-60. doi: 10.1400/233577

Pelgrom, J. & Stek, T.D. 2014. Roman Colonization under the Republic: historiographical contextualisation of a paradigm. In T.D. Stek & J. Pelgrom (eds.), *Roman Republican Colonization. New Perspectives from Archaeology and Ancient History*, 10-45. Rome: Palombi Editori.

Peterson, C.E. & Drennan, R.D. 2011. Patterned variation in regional trajectories of community growth. In M.E. Smith (ed.), *The Comparative Archaeology of complex societies*, 88-137. Cambridge: Cambridge University Press.

Piccarreta, F. & Ceraudo, G. 2000. *Manuale di aerofotografia archeologica. Metodologia, tecniche e applicazioni*. Bari: Edipuglia.

Plog, S., Plog, F. & Wait, W. 1978. Decision making in modern surveys. In M.B. Schiffer (ed.), Advances in archaeological method and theory 1, 383-421. New York: Academic Press.

Potter, T.W. 1979. *The Changing Landscape of South Etruria*. London: Elek.

Quilici, L. & Quilici Gigli, S. 2004. *Introduzione alla topografia antica*. Bologna: Il Mulino.

Rathbone, D.W. 1981. The development of agriculture in the 'Ager Cosanus' during the Roman Republic: Problems of evidence and interpretation. *The Journal of Roman Studies* 71, 10-23. doi: 10.2307/299493

Rathbone, D.W. 2008. Poor peasants and silent sherds. In L. De Ligt & S.J. Northwood (eds.), *People, land and politics. Demographic developments and the transformation of Roman Italy, 300 B.C.-A.D. 14*, 305-332. Leiden: Brill

Renfrew, C. & Bahn, P. 2008. *Archaeology: Theories, methods and practice*. Thames & Hudson.

Roberts, B. K. 1996. *Landscapes of settlement: prehistory to the present*. London: Routledge.

Salmon, E.T. 1969. *Roman Colonization under the Republic*. London: Thames and Hudson

Salmon, M. H. 1976. "Deductive" versus "Inductive" archaeology. *American Antiquity* 41(3), 376-381. doi:10.2307/279528

Schiffer, M. B. 1987. *Formation Processes of the Archaeological Record*. Albuquerque: University of New Mexico Press.

Settis, S. (ed.) 1984. *Misurare la terra: centuriazione e coloni nel mondo romano*. Modena: Edizioni Panini

Sharon, I., Dagan, Y. & Tzionit, G. 2004. The [awful?] truth about GIS and archaeology. In M. Iacovou (ed.), *Archaeological field survey in Cyprus: Past history, future potentials*, 151-62. London.

Sommella, P. 2009. Esperienze documentali sul territorio dagli anni '80 ad oggi. Alcune considerazioni. *Archeologia e Calcolatori* 20, 47-59.

Stek, T.D. 2009. *Cult places and cultural change in Republican Italy. A contextual approach to religious aspects of rural society after the Roman conquest*. Amsterdam: Amsterdam University Press.

Stek, T.D. 2014. The city-state model and Roman Republican colonization: sacred landscapes as a proxy for colonial socio-political organization. In T.D. Stek & J. Pelgrom (eds.), *Roman Republican Colonization. New Perspectives from Archaeology and Ancient history*, 87-105. Rome: Palombi Editori.

Stek, T.D. 2016. Less is more? A manifesto for targeted, problem-oriented research for Classical Archaeology students. In: F. Boschi (ed.), *Preventive Archaeology. Evaluating sites and landscapes*, 147-159. Bologna.

Stek, T. D. 2018. Early Roman colonization beyond the Romanizing agro-town: village patterns of settlement and highland exploitation in Central Italy. In B. S. Düring & T. D. Stek (eds.), *The Archaeology of Imperial Landscapes: A Comparative Study of Empires in the Ancient Near East and Mediterranean World*, 145-172. Cambridge: Cambridge University Press.

Stek, T. D., Modrall, E. B., Kalkers, R. A. A., van Otterloo, R. H. & Sevink, J. 2015. An early Roman colonial landscape in the Apennine mountains: Landscape archaeological research in the territory of Aesernia (Central-Southern Italy). *Analysis Archaeologica: An International Journal of Western Mediterranean Archaeology* 1, 229–291.

Stek, T. D. & Pelgrom, J. 2013. Landscapes of Early Roman Colonization: Non-urban settlement organization and Roman expansion in the Roman Republic (4th–1st centuries B.C.). *Tijdschrift voor Mediterrane Archeologie* 50, 87.

Stek, T.D., Pelgrom, J., Casarotto, A., García Sánchez, J., Götz, L., Hamel, A., Iannantuono, K., Kalkers, R.A.A., Termeer, M.K. & Waagen, J. 2016. Santuari, villaggi, centri fortificati e prima urbanizzazione tra Sanniti e Romani: Nuove ricerche archeologiche nell'Alta Valle del Tappino, nel territorio della colonia latina di Aesernia e in quello della città frentana di Larinum. *ArcheoMolise* 26. 24-39.

Stone, D. L. 1997. *The development of an Imperial territory: Romans, Africans and the transformation of the rural landscape of Tunisia*. PhD thesis, University of Michigan: University of Michigan Library.

Tashakkori, A. & Teddlie, C. 2003. *Handbook of mixed methods in social & behavioral research*. Thousand Oaks: Sage.

Terrenato, N. & Ammerman, A. J. 1996. Visibility and site recovery in the Cecina valley survey, Italy. *Journal of Field Archaeology* 23 (1), 91-109. doi: 10.1179/009346996791973990

Thompson, S. 2004. Side-by-side and back-to-front: exploring intra-regional latitudinal and longitudinal comparability in survey data. Three case studies from Metaponto, southern Italy. In S.E. Alcock & J.F. Cherry (eds.), *Side-by-side survey: Comparative regional studies in the Mediterranean world*, 65-85. Oxford: Oxbow.

Torri, D., Borselli, L.,Guzzetti, F., Costanza Calzolari, M., Bazzoffi, P., Ungaro, F., Bartolini, D. & Pilar Salvador Sanchis, M. 2006. Italy. In J.

Boardman & J. Poesen (eds.), *Soil erosion in Europe*, 245-261. Chichester, West Sussex, England: John Willey & Sons, Ltd.

Traviglia, A. 2016. Beyond the city walls: Surveying the periphery of Aquileia. In *Australian Archaeological Fieldwork Abroad III. A special issue of 'Ancient History: Resources for Teachers'* 45, 64-94.

Van Leusen, P. M. 2002. *Pattern to process: Methodological investigations into the formation and interpretation of spatial patterns in archaeological landscapes*. PhD dissertation, University of Groningen: University of Groningen Research Database. http://hdl.handle.net/11370/fc9ead23-e2c7-40c5-942f-e06f52198222

Van Oost, K. & Govers, G. 2006. Tillage erosion. In J. Boardman & J. Poesen (eds.), *Soil erosion in Europe*, 599-608. Chichester, West Sussex, England: John Willey & Sons, Ltd.

Verhagen, P. 2007. *Case studies in archaeological predictive modelling* (ASLU 14). Leiden: Leiden University Press.

Verhagen, P. 2013. Site discovery and evaluation through minimal interventions: core sampling, test-pits and trial trenches. In C. Corsi, B. Slapšak & F. Vermeulen (eds.), *Good practice in archaeological diagnostics: Non-invasive survey of complex archaeological sites*, 209-225. Springer

Verhagen, P. & Whitley T. G. 2012. Integrating archaeological theory and predictive modelling: A live report from the scene. *Journal of Archaeological Method and Theory* 19 (1), 49 – 100. doi: 10.1007/s10816-011-9102-7

Vermeulen, F., Van Limbergen, D., Monsieur, P. & Taelman, D. 2017. *The Potenza Valley Survey (Marche, Italy). Settlement dynamics and changing material culture in an Adriatic valley between Iron Age and Late Antiquity*, Academia Belgica. Studia Archaeologica 1. Rome: System Graphic Srl.

Ward-Perkins, J. B. 1961. Veii: The Historical Topography of the Ancient City. *Papers of the British School at Rome* 29, 1–119. DOI: https://doi.org/10.1017/S0068246200010916

Wheatley, D. 2004. Making space for an archaeology of place. *Internet archaeology* 15. doi: 10.11141/ia.15.10

Wheatley, D. & Gillings, M. 2002. *Spatial technology and archaeology. The archaeological applications of GIS*. London & New York: Taylor & Francis.

Wilkinson, T. J. 2000. Regional approaches to Mesopotamian archaeology: The contribution of archaeological surveys. *Journal of Archaeological Research* 8 (3), 219-267.

Wilkinson, T.J. 2003. *Archaeological landscapes of the Near East*. The Arizona Board of Regents: The University of Arizona Press.

Wilson, A. 2008. Site recovery rates and the ancient population of the Biferno Valley. In G. Lock & A. Faustoferri (eds.), *Archaeology and landscape in Central Italy: papers in memory of John A. Lloyd*, 233-253. Oxford: School of Archaeology, University of Oxford.

Wise, A. & Miller, P. 1997. Why metadata matters in archaeology. *Internet archaeology* 2. doi:10.11141/ia.2.5

Witcher, R. E. 2005. The extended metropolis: Urbs, suburbium and population. *Journal of Roman archaeology* 18, 120-138.

Witcher, R. E. 2008. (Re)surveying Mediterranean rural landscapes: GIS and legacy survey data. *Internet Archaeology* 24. doi: 10.11141/ia.24.2.

Witcher, R.E. & Craven, M. 2013. 'Much that has long been hidden': Reconstructing the survey methodology. In H. Patterson, H. Di Giuseppe & R. Cascino (eds.), *Veii. The historical topography of the ancient city. A restudy of John Ward-Perkins' survey*, 9-24. London: British School at Rome.

WWF, 2017. *Caring for our soil. Avere cura della natura dei territori*. WWF Report I 2017. Rome: WWF Italia Ong Onlus

Yntema, D. G. 1993. In search of an ancient countryside: The Amsterdam Free University field survey at Oria, province of Brindisi, south Italy (1981–1983). (Scrinium; No. VI). Amsterdam, Thesis Publishers.

Chapter 2. A method for modeling dispersed settlements: visualizing an early Roman colonial landscape as expected by conventional theory[9]

Abstract

This paper proposes a GIS quantitative method for simulating dispersed distributions of sites in a landscape. A number of sites might have escaped archaeological detection due to adverse surface visibility conditions during field survey (the so-called 'missing sites'). For the early Roman colonial landscapes of central-southern Italy, these surface visibility factors are traditionally seen to be so dramatic as to have allegedly hampered the detection of the conventionally expected dispersed and densely-settled colonial farm landscape. In this paper, the regional and site-oriented field survey conducted in Venosa (Basilicata, Italy) is used as a case-study to simulate hypothetical early colonial sites. The aim of this theoretical exercise is to show how the rural dispersed settlement pattern expected by the conventional theory could appear on a map, and to visually highlight the divergence between survey data and conventional spatial expectancies.

2.1 INTRODUCTION

In the debate on ancient settlement organization of Roman colonies, the extent to which rural colonial settlements have been detected by modern archaeologists in the field plays an important role. A loss in recovery of sites as a result of the limitations of field survey methodology (*i.e.* visibility biasing factors) is generally known (see discussion in De Guio 1985; 1995; Francovich *et al.* 2000), but its precise scale is seldom clearly connected to historical interpretation. In the conventional reconstruction of Roman colonial landscapes, this notion of a low site recovery rate is crucial to the argumentation. In this paper, the colonial landscape projected by the conventional view, and by extension the related levels of site 'loss' between the Roman period and modern day field observations, is calculated and visualized by means of a GIS simulation. This not only reveals the scale of modern correction needed to sustain the conventional view of early Roman landscapes, but also introduces a useful visualization method.

Numerous studies have addressed the impact of factors that could prevent object detection in the plough-soil (*i.e.* artifact taphonomy). Several methods have been proposed for the correction of artifact density recorded in intensive, off-site field survey (*e.g.* Haselgrove *et al.* 1985; Shennan *et al.* 1985; Allen 1991; Gaffney *et al.* 1991; Schofield 1991; Verhoeven 1991; Van de Velde 1996; 2001; van Leusen 1996; 2002; Gillings & Sbonias 1999; Fentress 2000; Ebert & Singer 2004; Given 2004a and b; De Haas 2012; Feiken 2014; Waagen 2014). In contrast, the development of methods to "correct" site recovery rates in site-based field surveys has seen less research (discussion in Cambi 1999; Witcher 2006; 2011; Wilson 2008; Fentress 2009). For

9 This Chapter corresponds to the article "A method for modeling dispersed settlements: visualizing an early Roman colonial landscape as expected by conventional theory" by Anita Casarotto, originally published (22 December 2017) in *Archeologia e Calcolatori*, volume 28.1 (publisher Edizioni all'Insegna del Giglio) as an Open Access article distributed under the terms of the Creative Commons Attribution-NonCommercial-NoDerivatives License (https://creativecommons.org/licenses/by-nc-nd/3.0/). This article is accessible through this link: https://doi.org/10.19282/AC.28.1.2017.09
I would like to express my gratitude to my supervisors, Dr. Jeremia Pelgrom and Dr. Tesse D. Stek for their comments throughout the writing process of the article. I would also like to thank Prof. Maria Luisa Marchi for the methodological information about the visibility map of the territory of Venusia. Limited changes to the original publication have been included in this Chapter 2.

surveys in Italy, only a few studies have proposed concrete methods for simulating the sites that may have not been detected due to poor ground visibility conditions in which site-based surveys were carried out (*e.g.* Terrenato & Ammerman 1996; Terrenato 2000; see also Nance 1983 for a methodological discussion and Fokkens 1998 for a north-western European case-study). In this paper these previous methods as well as new ones are taken into account to simulate the possible effect of surface visibility on site detection in survey.

Following the conventional theory on colonial settlement (see below), hypothetically "corrected" site recovery rates and "complete" site distributions are, by way of experiment, proposed for the Roman colonial landscape of the Latin colony of Venusia (founded in 291 B.C. in southern Italy). The field surveys conducted in the territory of Venusia, and in many other rural colonial landscapes around the Italian peninsula, identified only a fraction of the early colonial sites (and specifically only a small portion of the 3rd century B.C. farms) expected based on the demographic information reported in ancient literary sources (see the discussion in Rathbone 1981; 2008; Pelgrom 2008; 2013). The generally accepted explanation for this *conundrum* of missing colonial sites is that they have not been identified because of the difficulty to detect simple, poor colonists' rural dwellings in pedestrian surveys, especially when the surface visibility conditions are not optimal for site detection (Rathbone 1981; 2008; Millett 1991; Cambi 1999; Witcher 2011). According to this conventional idea, this obstacle to site discovery would explain why the expected densely populated and regularly settled colonial landscape is not visible through field survey.

Using a quantitative method for correcting possible survey visibility distortions in settlement patterns, this study shows how the territory around the colonial center of Venusia may have appeared if the conventional model is correct (for this model of dispersed Roman colonial settlements see *e.g.* Salmon 1969; Brown 1980; Rathbone 1981; 2008; Celuzza & Regoli 1982; Settis 1984; Bussi & Vandelli 1985; for a critique of this model see Pelgrom 2008; 2013; Stek & Pelgrom 2013). This was achieved by means of computer-based simulations that uniformly allocated a large amount of hypothetical missing sites (in other words, sites that might not have been registered by previous field surveys because of the less than ideal ground visibility conditions). The result is a hypothetical reconstruction of the expected "complete" site distribution (supposedly the distribution that, according to the conventional model, should have been observed during the survey if surface visibility was constantly, not variably, optimal across the entire landscape). The assumption underlying such simulations of landscapes with dispersed settlements is that there is a strong association between surface visibility and the number of colonial sites discovered in field surveys; an idea that, as previously mentioned, is widely accepted by the conventional theory on colonial settlement.

Prior calculations of site recovery rates for colonial landscapes have been based upon preconceptions about rural colonial landscapes, and which recovery rates we would end up with when following certain established assumptions on density and distribution (*e.g.* Rathbone 1981; Cambi 1999). This analysis further explores this line of thought by offering a visual, concrete picture of how the envisaged dispersed colonial settlement pattern would have been recorded, if it existed and was visible, on a site-based survey map. This theoretical exercise serves as an important first step to gaining a better understanding of the quantitative and spatial implications of this conventional theory. However, the alleged direct relationship between object of study (*i.e.* distribution and density of sites in colonial landscapes) and methodological survey limitations (*i.e.* visibility biasing factors) cannot be taken at face value but needs to be tested (see Chapter 5).

Recently, several scholars have started to question the conventional Roman colonial settlement theory, which predicts a colonial countryside settled regularly and densely. By noting that, instead, irregular patterns underlie the settlement sites registered in surveys, the question is now raised as to whether these patterns reflect another, alternative settlement model (Pelgrom 2008; Stek 2009) rather than being the result of visibility biasing factors (Rathbone 2008). The current paper aims to contribute to this

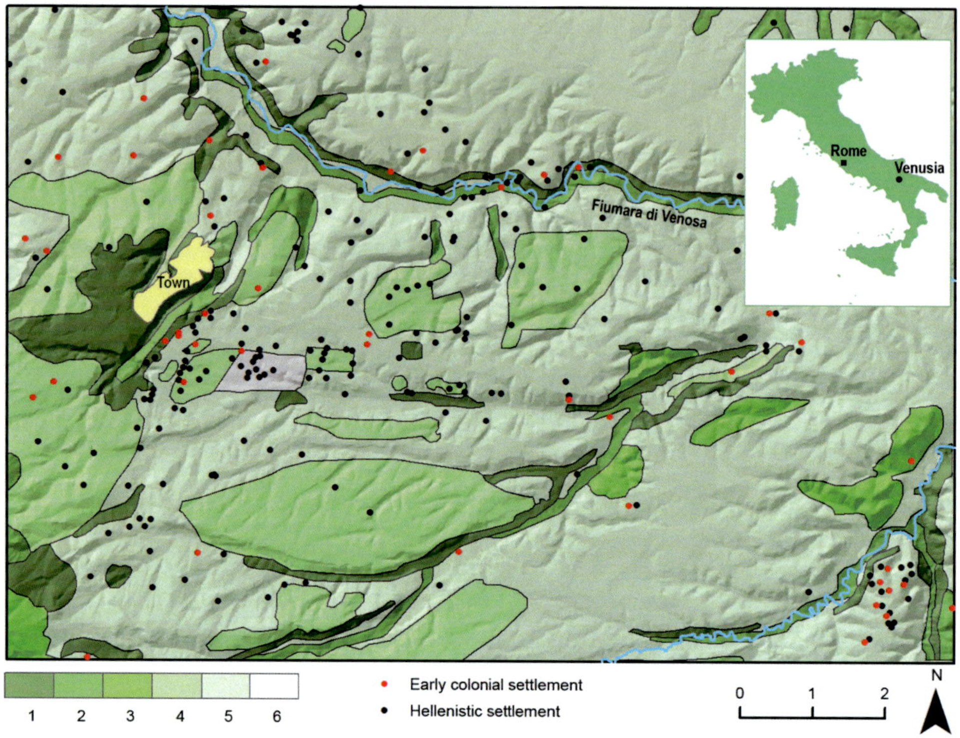

Fig. 2.1 Hellenistic settlement site distribution and visibility map (based on Marchi & Sabbatini 1996, 107; graphic elaboration by author). The raster base map is the shaded relief calculated from the 10 m-resolution DEM named TINITALY/01 (Tarquini *et al.* 2007; 2012; Tarquini & Nannipieri 2017).

debate by offering a tool that enables a visual, direct assessment of the substantial spatial discrepancy that exists between the conventional settlement theory and survey data (see discussion in Pelgrom 2012). This, in turn, can be used to evaluate the validity of this model for the settlement organization of the colonial countryside.

2.2 DATA

The regional, site-oriented field survey in north-eastern Basilicata was conducted between 1989 and 2000 within the context of the *Forma Italiae* project (Azzena & Tascio 1996; Sommella 2009; Marchi 2016a and b; Marchi & Forte 2016), and represents one of the richest datasets of this type for central-southern Italy (Marchi & Sabbatini 1996; Sabbatini 2001; Marchi 2010)[10]. The first data published were collected in the area around the ancient urban center of the Latin colony of Venusia (*c.* 120 sq km) (Marchi & Sabbatini 1996). In total 604 sites were detected here, of which 262 are settlement sites dating to the Hellenistic period. Of these 262 sites, there are 44 settlement sites dated with certainty to the early colonial period, given that 3rd century black gloss pottery was attested at these sites (Fig. 2.1).

In this paper, all Hellenistic settlement sites are taken into consideration as possible early colonial settlements. This is because sites possibly occupied in the 3rd century B.C. but lacking datable diagnostic

10 Fields were systematically walked by surveyors spaced five to ten m apart who recorded all visible material concentrations (*i.e.* sites) with a material density equal to or higher than five shards per sq m (see Pelgrom *et al.* 2014, 31-35; Marchi 2016a for more details on the survey method). IGM (Istituto Geografico Militare) maps (1:25,000) were used as a support to register the position and the extension of sites, in combination with CTRs maps (Carta Tecnica Regionale 1:5000; 1:10,000), and GPS technologies (Azzena & Tascio 1996).

archaeological evidence (*e.g.* 3rd century B.C. black gloss pottery) can be dated only to a broader chronological range (namely the Hellenistic period) (for further argumentation of this choice see also Chapter 3, section 3.2). In the current simulation, settlement sites are treated as simple and homogenous dots[11].

The survey team paid attention to the relationship between ground visibility and site discovery. After noting that different land uses and the plow status of the fields offered suitable visibility for survey especially in certain seasons (*e.g.* olive and vineyard orchards in winter and spring), repeated coverage of the same surface was planned and executed in order to retrieve the necessary information. For the territory around the ancient town of Venusia, a visibility map was produced by the team and published in Marchi & Sabbatini 1996, 107. A graphic elaboration of this map (by author) is offered in Fig. 2.1.

Such a map integrates the land use information[12] and the land cover conditions registered by field walkers in the season of optimal visibility for survey, and classifies the land surfaces on a visibility scale from one (null or very low visibility) to six (optimal visibility) (Azzena & Tascio1996, 292-296). With this map, surveyors wanted to test if there is a link between the low number of archaeological sites recorded in wide portions of the landscape and surface visibility. The archaeological site points displayed in their distribution maps had a particular pattern: very densely populated zones constantly alternated with much less dense site areas and large vacuums (Marchi & Sabbatini 1996, 103-104, 111-130). The visibility map can help to evaluate whether such a configuration of sites is the result of visibility biasing factors or, rather, the result of ancient settlement rationales.

11 A majority of these sites was occupied during several phases. Therefore, their size is not necessarily indicative of an early colonial occupation and may likely be related to those archaeologically "more visible" periods characterized by the abundant consumption of non-perishable material. As a consequence, it would be incorrect to use the documented size as a parameter for distinguishing different settlement categories (*e.g.* farm, villa, village) for individual periods. The aim here is not to perform analyses of visibility for correcting the number of different settlement categories. The fact that there exists a well-acknowledged problematic relationship between material concentrations recorded at the surface during surveys and the reliability of site classifications based on this type of data (*e.g.* Barker & Lloyd 1991; Alcock & Cherry 2004) justifies here the methodological choice of disregarding categories in favor of a more neutral definition of surface material scatters.

12 For more details see Azzena & Tascio 1996, 292 - 296 (especially footnote 18).

2.3 VISUALIZING THE CONVENTIONALLY EXPECTED EARLY COLONIAL LANDSCAPE

In this section, a quantitative method is proposed to correct possible visibility distortions in site density and distribution. A similar method has been presented by Terrenato (2000). His approach also aimed at measuring the probability of a site being present in a certain location, even though it had not been identified by surveyors due to limited visibility. The relation between local variations in ground visibility and site discovery is complex; the variable effect of visibility on site recovery rates must be taken into account when constructing a simulation of missing sites. Additionally, the simulation of a dense and dispersed early colonial landscape is based on the assumption of uniformity in the original settlement distribution. Therefore, hypothetical colonial missing sites should be allocated across the landscape in a way that is calibrated in accordance with the relationship between visibility and the recorded site density and pattern at each landscape location. This simulation was conducted in three steps.

As a first step, a probability surface for the allocation of hypothetical missing sites was constructed. This surface must indicate the likelihood that a missing colonial site existed in each location of the landscape, but was not recorded by surveyors. This probability thus depends on both ground visibility conditions experienced during the survey, and site density recorded at each landscape location. The calculation of this surface was implemented in ArcGIS 10.2.2 using three GIS tools:

1) First, a kernel density estimation of Hellenistic sites was carried out (Fig. 2.2). The kernel density tool of ArcGIS calculates "a smooth estimate of a probability density from an observed sample of observations" (Bailey & Gatrell 1995, 84). The probability density is highest at the location of points and diminishes gradually with increasing distance from the points

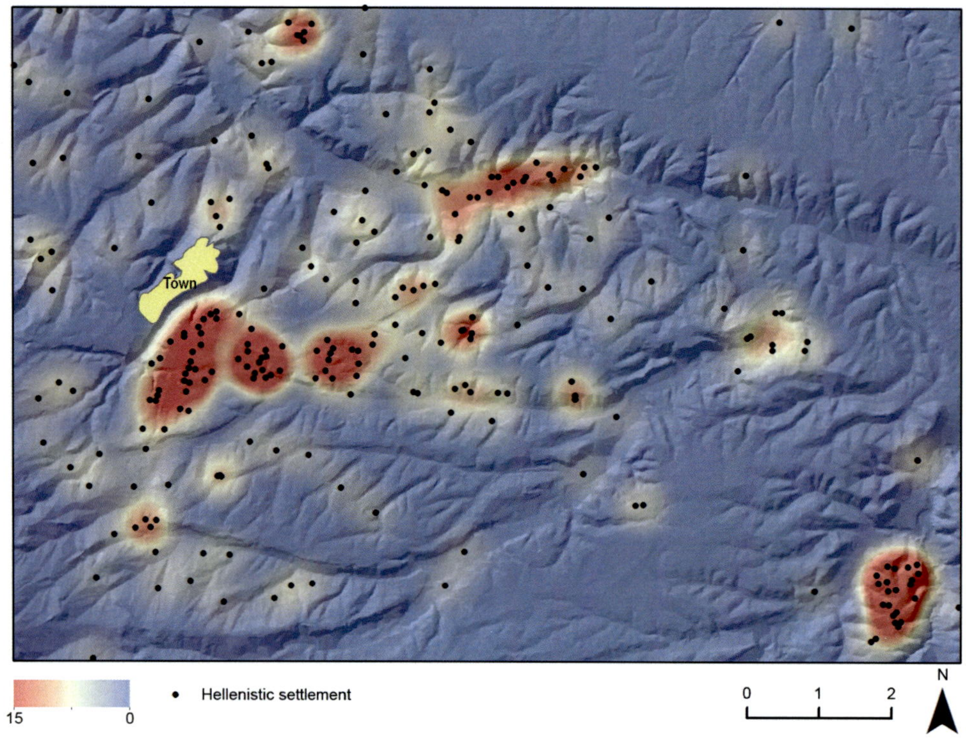

Fig. 2.2 Kernel density surface calculated for the Hellenistic settlements. The legend indicates the number of estimated sites in a circle of one square km from each location (cell resolution 10 x 10 m). The raster base map is the shaded relief calculated from the 10 m-resolution DEM named TINITALY/01 (Tarquini *et al.* 2007; 2012; Tarquini & Nannipieri 2017).

(Esri 2014a). The site probabilities were calculated in a circle of one sq km from each recorded point, and then were summed up in each location (cell size set at ten by ten m). The resulting density surface displayed the probable intensity for a particular distributional phenomenon in each location and over the entire landscape, and is based on the density of recorded Hellenistic sites.

2) Afterwards, the fuzzy tool of ArcGIS was used to linearly and inversely transform the kernel density estimate (Fig. 2.2) and the visibility map (Fig. 2.1) into probability surfaces, with values that range from zero to one (Esri 2014b). The traditional model for the organization of the colonial countryside expects high and homogeneous site density; following this idea, landscape locations with a low recorded site density received a high probability value for missing colonial sites; conversely, landscape locations with a high recorded site density received a low probability value for missing sites. This means that, for instance, the maximum site density value had the lowest probability (*i.e.* zero) and the minimum density value had the highest probability (*i.e.* one) for missing sites. The same operation was performed for the ground visibility map: high probability values for missing colonial sites were given to low values of visibility, and low probability values for missing sites were assigned to high values of visibility. Again, this means that the maximum and the minimum visibility values (*i.e.* six and one, see Fig. 2.1) received respectively the lowest and the highest probability for missing sites (*i.e.* zero and one).

3) The reclassification of these two variables (*i.e.* site density and ground visibility) on the same probability scale (from zero to one) made them unitless, and allowed for direct comparison and integration. Fig. 2.3 shows the result of the integration of these two variables. In order to correct the estimated site density for possible visibility biases, the reclassified density map was multiplied by the reclassified visibility map

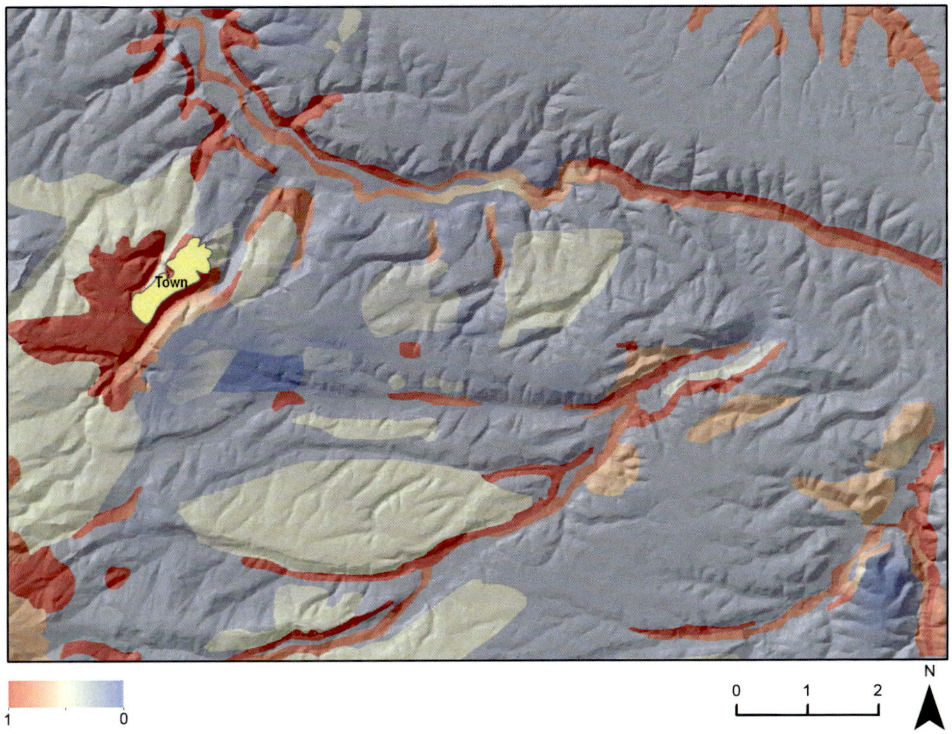

Fig. 2.3 Trend surface created for missing site allocation. The legend indicates the probability for the allocation of missing sites and ranges from a minimum of 0 to a maximum of 1. The raster base map is the shaded relief calculated from the 10 m-resolution DEM named TINITALY/01 (Tarquini *et al.* 2007; 2012; Tarquini & Nannipieri 2017).

through a raster overlay operation. In this way, a trend surface for the allocation of missing sites was created. This trend surface takes into consideration the possible visibility distortions involved in the recording of sites and can be used as a base to allocate hypothetical missing sites across the landscape.

As a second step, it was necessary to calculate the hypothetical number of missing sites, before implementing the allocation of points on the trend surface. There are already figures for missing sites in colonial landscapes that were calculated by scholars following a demographic method (see below). In this analysis, however, these previously-proposed numbers were not used; instead, new thresholds for missing sites were calculated using a GIS-based spatial and statistical method. Specifically, the number of sites necessary to recreate a dispersed pattern was calculated through several tests: increasingly larger numbers of missing sites were simulated until the site distribution clearly started to exhibit a statistically significant regular pattern. The pattern of the simulated distributions was evaluated by means of the nearest neighbor tool[13] of ArcGIS (for other methods see *e.g.* Orton 2004; Ducke 2015).

As a third step, the trend surface was used as a base for the allocation of simulated sites. The allocation of

13 This tool allows researchers to categorize the dominant pattern displayed by the simulated dots (either clustered, random, or dispersed) (Clark & Evans 1954). The nearest neighbor analysis calculates the distances from each point to its nearest neighbor and then averages all these inter-distances (Hodder & Orton 1976, 30-52; Kintigh & Ammerman 1982; see the discussion in Orton 2004). If this average is higher than that obtained from a random distribution of dots, the site distribution exhibits regularity (*i.e.* dispersed distribution). This tool also calculates the nearest neighbor ratio by dividing the observed average distance by the expected average distance: a ratio less than 1 indicates clustering, equal to 1 randomness, and more than 1 indicates uniformity (Esri 2014c). The z-score and the p-value resulting from using this tool, then, reveal whether the detected pattern is significant: in other words, a significant pattern is identified if very high or very small z-scores exist in association with very small p-values (see Esri 2014c for the mathematical details on this procedure).

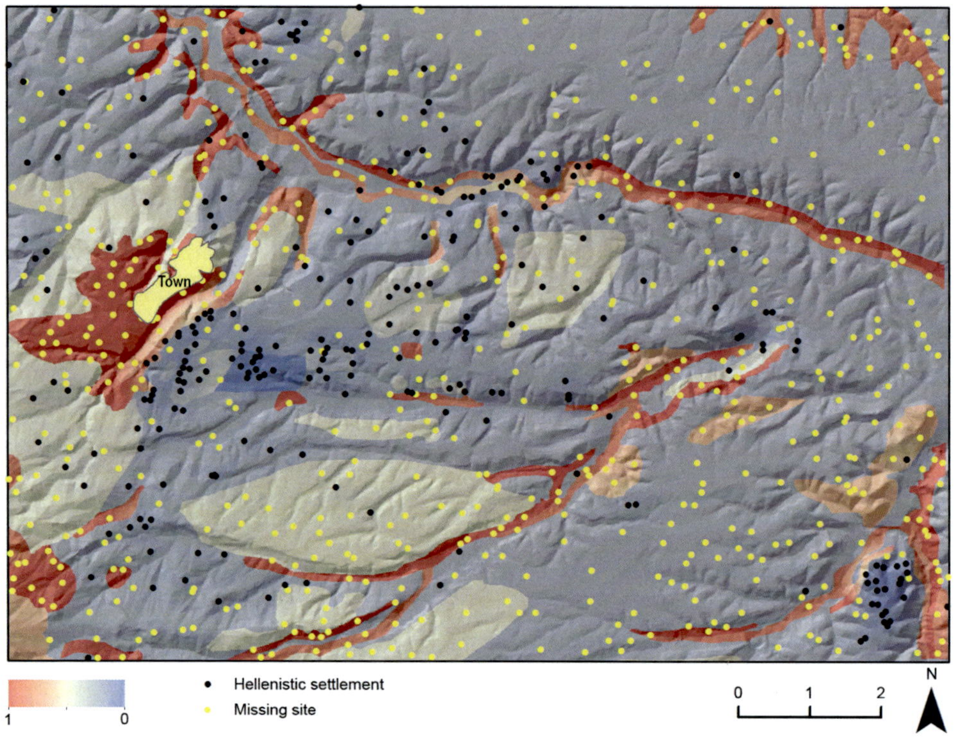

Fig. 2.4 Trend surface created for the allocation of missing sites (probability from 0 to 1), and the simulated "complete" early colonial settlement distribution (262 recorded settlements plus 600 missing sites, therefore 862 sites in total). The raster base map is the shaded relief calculated from the 10 m-resolution DEM named TINITALY/01 (Tarquini *et al.* 2007; 2012; Tarquini & Nannipieri 2017).

missing sites on the trend map (Fig. 2.4) is based on the conventional assumption of homogeneous site density and pattern for the colonial countryside. Only few sites (or none) were thus allocated in those landscape locations where the recorded site density was high and the ground visibility was good, since the probability for missing sites there was low. As the density and visibility decreased, more sites were allocated: in the zones where both the recorded site density and visibility were low, many sites were spread out because of the high probability for missing sites in these locations. In other words, those landscape locations characterized by both low visibility and low recorded site density were more likely to receive missing sites than locations with a high recorded site density in good visibility conditions. In ArcGIS there is a tool precisely suited for this operation: it is called 'create spatially balanced points'. By using this GIS tool, points representing missing sites were scattered across the study area in a balanced way (*i.e.* in proportional accordance with the trend surface probability values, see above) to create a uniformly dotted landscape. As a future step to improve this simulation, it would be interesting to consider other constraints for the allocation of sites (*e.g.* the relief), and to note to what extent the exclusion of, for instance, steep slopes changes the simulated site configuration.

By using the nearest neighbor tool, the pattern type characterizing the simulated distributions was assessed (see footnote 13, p. 35). Thus, the number of missing sites required to create a dispersed early colonial settlement distribution was established. For our case-study, the originally recorded Hellenistic distribution (262 sites in total) exhibited clustering in its pattern (neighbor ratio: 0.707420; z-score: -9.059943; p-value: 0.00). In order to transform this clustered distribution into a regularly dense distribution of early colonial sites, a total number of (at least) 600 missing sites had

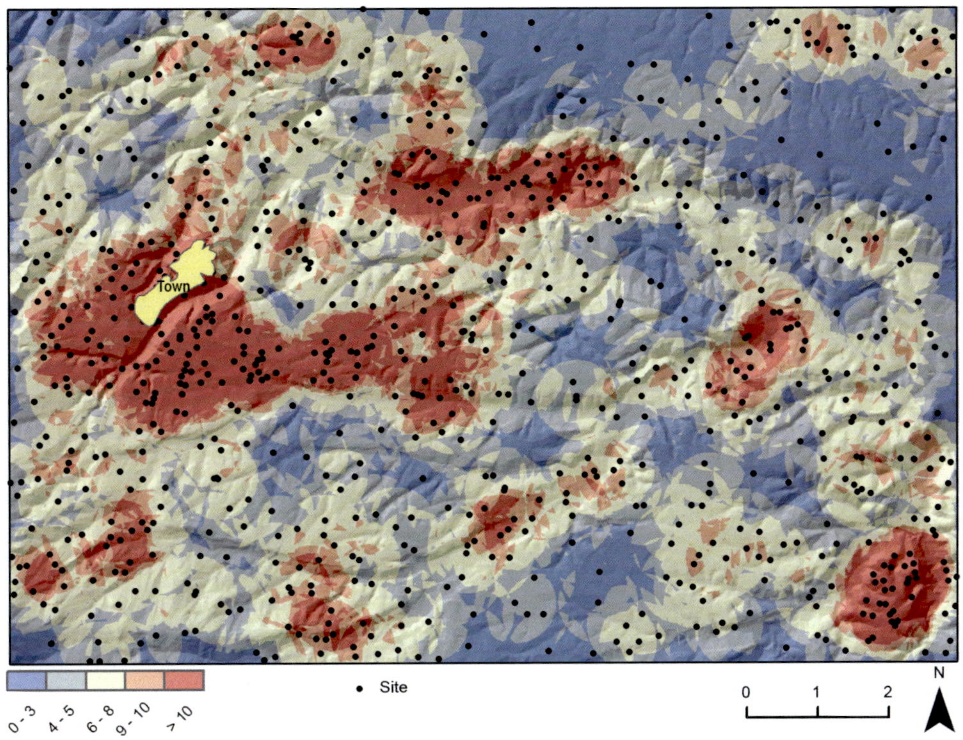

Fig. 2.5 Point-density analysis of the "complete" early colonial distribution (262 recorded sites plus 600 missing sites, therefore 862 sites in total). The legend indicates the number of sites located in a circle of one square km from each location (cell resolution 10 x 10 m). The raster base map is the shaded relief calculated from the 10 m-resolution DEM named TINITALY/01 (Tarquini *et al.* 2007; 2012; Tarquini & Nannipieri 2017).

to be allocated over the trend surface and added to the 262 recorded settlements (Fig. 2.4). From this new hypothetical "complete" distribution (in total 862 sites), the nearest neighbor ratio tallies to 1.037621, the z-score is 2.113060, and the p-value is 0.034596.

This simulated dispersed settlement distribution has an average density of 6.8 sites per sq km (Fig. 2.5). What is interesting to note is that this figure aligns very well with the demographic information cited in the literary sources, from which a number of 7-8 colonists' farms per sq km has been calculated by previous studies (Pelgrom 2008; 2012; 2013). Moreover, this amount of missing colonial sites is consistent with the 20-33% recovery rate that has been estimated for a regional site-based survey carried out in another Roman colonial context in Italy (*i.e.* the Albegna valley in Tuscany for the Latin colony of Cosa, Cambi 1999; for other calculations see Rathbone 1981; Wilson 2008; Witcher 2011). Indeed, for this simulated early colonial distribution (Fig. 2.4), the number of hypothetical unrecorded missing sites amounts to 69.6% of the total. Therefore, the number of recorded sites corresponds to 30.4% of the total. Interestingly, the GIS method described in this paper proposes thresholds for both the expected and the missing colonial sites that comply with the thresholds calculated by other scholars who used a completely different method (*i.e.* text-based demographic method, cf. *supra*).

It is important to stress, however, that the scale of the source visibility map played a key role in the simulation. If finer, more-detailed base maps on which to perform the allocation of missing sites were available, a different hypothetical distribution of early colonial settlements would probably have been obtained, and thus different percentages for recorded and missing sites.

2.4 DISCUSSION AND FURTHER DIRECTIONS

It is generally understood by archaeologists that biasing factors such as ground visibility conditions can strongly affect the detection of sites. These potential biases are usually seen as the explanation for the missing colonial farm sites in most field survey projects in central-southern Italy (*e.g.* Rathbone 2008). The legacy site distribution maps compiled during these projects do not show, as a rule, the expected evenly dotted colonial landscape.

In this paper, a theoretical exercise to spatially visualize the conventionally expected site configuration was offered. If we accept such an understanding of colonial settlement organization, as well as the existence of a strong relationship between site density and survey visibility, the computer-based simulation proposed here clearly shows the high number of sites that need to be simulated in order to create a dispersed distribution.

However, this conclusion of very high missing site percentages fully depends on the validity of the conventional and historically informed theories about colonial population density and settlement configuration. As this understanding of colonial rural landscapes has recently been undermined by a series of revisionist studies, the question arises whether the idea of biasing factors heavily hampering the detection of this alleged dispersed landscape is actually correct (see Chapters 4 and 5). Therefore, as a subsequent research step, the validity of the conventional model needs to be tested using descriptive methods and computer-based statistical tools, as well as in the field with new surveys.

As a matter of fact, the new colonial settlement models that have been proposed recently not only question the conventional theories on Roman territorial strategies in recently conquered areas, but also have a significant impact on broader discussions about missing sites and field survey biases (Pelgrom 2008; Stek & Pelgrom 2013; Stek *et al.* 2015; Casarotto *et al.* 2016; Casarotto *et al.* 2018; for a discussion of the scholarly debate related to these settlement models see *e.g.* Bispham 2006; Bradley 2006; Terrenato 2007; Van Dommelen & Terrenato 2007; Stek & Pelgrom 2014). The focus now shifts from showing how sites could have appeared on a map to understanding whether ancient settlements should even be expected in the first place, and most importantly, which site patterns are the result of biasing factors, and which, instead, are the result of genuine settlement preferences.

2.5 REFERENCES

Alcock, S.E. & Cherry, J.F. (eds.) 2004. *Side-by-side survey: Comparative regional studies in the Mediterranean world.* Oxford: Oxbow Books.

Allen, M.J. 1991. Analysing the landscape: A geographical approach to archaeological problems. In A.J. Schofield (ed.), *Interpreting artefact scatters: Contributions to ploughzone archaeology*, 39-57. Oxford: Oxbow Monograph 4.

Azzena G., & Tascio, M. 1996. Il Sistema Informativo Territoriale per la carta archeologica d'Italia. In M.L. Marchi & G. Sabbatini, *Venusia (IGM 187 I NO / I NE)* (Forma Italiae 37), 281 – 297. Firenze: Leo S. Olschki.

Bailey, T.C. & Gatrell A.C. 1995. *Interactive spatial data analysis*. Harlow, Essex: Longman Scientific & Technical.

Barker, G. & Lloyd, J. (eds.) 1991. *Roman landscapes: Archaeological survey in the Mediterranean region.* London: British School at Rome.

Bispham, E. 2006. Coloniam deducere: How Roman was Roman colonization during the Middle Republic? In G. Bradley & J.P. Wilson (eds.), *Greek and Roman colonization: Origins, ideologies and interactions*, 73-160. Swansea: Classical Press of Wales.

Bradley, G. 2006. Colonization and identity in Republican Italy. In G. Bradley & J.P. Wilson (eds.), *Greek and Roman Colonization: Origins, ideologies and interactions*, 161-187. Swansea: Classical Press of Wales.

Brown, F.E. 1980. *Cosa: The making of a Roman town*. Ann Arbor: University of Michigan Press.

Bussi, R. & Vandelli, V. (eds.) 1985. *Misurare la terra: centuriazione e coloni nel mondo Romano. Città, agricoltura, commercio: materiali da Roma e dal suburbio*. Modena: Edizioni Panini.

Cambi, F. 1999. Demography and Romanization in Central Italy. In J.L. Bintliff & K. Sbonias (eds.), *Reconstructing past population trends in Mediterranean Europe (3000 B.C.-A.D. 1800)*, 115-129. Oxford: Oxbow Books.

Casarotto, A., Pelgrom, J. & Stek, T.D. 2016. Testing settlement models in the early Roman colonial landscapes of Venusia (291 B.C.), Cosa (273 B.C.) and Aesernia (263 B.C.). *Journal of Field Archaeology* 41 (5), 568-586. doi:10.1080/00934690.2016.1211474

Casarotto, A., Stek, T.D., Pelgrom, J., van Otterloo, R.H. & Sevink, J. 2018. Assessing visibility and geomorphological biases in regional field surveys: The case of Roman Aesernia. *Geoarchaeology: An International Journal* 33 (2): 177 – 192. doi: 10.1002/gea.21627

Celuzza, M.G. & Regoli, E. 1982. La Valle d'Oro nel territorio di Cosa. Ager Cosanus e Ager Veientanus a confronto. *Dialoghi di Archeologia* 4, 31-62.

Clark, P.J. & Evans, F.C. 1954. Distance to nearest neighbour as a measure of spatial relationships in populations. *Ecology* 35, 444-453. doi: 10.2307/1931034

De Guio, A. 1985. Archeologia di superficie ed archeologia superficiale. *Quaderni di Archeologia del Veneto* 1, 176-184.

De Guio, A. 1995. Surface and subsurface: Deep ploughing into complexity. In W. Hensel, S. Tabaczynski & P. Urbanczyk (eds.), *Theory and practice of archaeological research*, II, 329 – 414. Warszawa: Institute of Archaeology and Ethnology, Commiteee of Pre- and Protohistoric Sciences, Polish Academy of Sciences.

De Haas, T.C.A. 2012. Beyond dots on the map: Intensive survey data and the interpretation of small sites and off-site distributions. In P. Attema & G. Schörner (eds.), *Comparative issues in the archaeology of the Roman rural landscape: Site classification between survey, excavation and historical categories*, 55-79. Portsmouth: JRA Supplementary Series 88.

Ducke, B. 2015. Spatial cluster detection in archaeology: Current theory and practice. In J.A. Barcelo and I. Bogdanovic (eds.), *Mathematics and Archaeology*, 352 – 368. Boco Raton: CRC Press.

Ebert, D. & Singer M. 2004. GIS, predictive modelling, erosion, site monitoring. *Assemblage* 8.

Esri, 2014a. Kernel Density (Spatial Analyst). In *ArcGIS 10.2.2 Help*.

Esri, 2014b. Fuzzy Memebership (Spatial Analyst). In *ArcGIS 10.2.2 Help*.

Esri, 2014c. Average Nearest Neighbor (Spatial Statistics). In *ArcGIS 10.2.2 Help*.

Feiken, M. 2014. *Dealing with biases: Three geo-archaeological approaches to the hidden landscapes of Italy*. Eelde: Barkhuis.

Fentress, E. 2000. What are we counting for? In R. Francovich, H. Patterson & G. Barker 2000 (eds.), *Extracting meaning from plough soil assemblages*, 45-52. Oxford: Oxbow Books.

Fentress, E. 2009. Peopling the countryside: Roman demography in the Albegna valley and Jerba. In A. Bowman & A. Wilson (eds.), *Quantifying the Roman economy: Methods and problems*, 127-163. Oxford: Oxford University Press.

Fokkens, H. 1998. *Drowned landscape. The occupation of the Western part of the Frisian-Drentian plateau, 4400 B.C.-A.D. 500*. Assen, Amersfoort, Van Gorcum: ROB, Rijksdienst Voor Het Oudheidkundig Bodemonderzoek.

Francovich, R., Patterson H. & Barker, G. (eds.) 2000. *Extracting meaning from plough soil assemblages*. Oxford: Oxbow Books.

Gaffney, V.L., Bintliff, J.L. & Slapsak B.1991. Site formation processes and the Hvar survey project, Yugoslavia. In A.J. Schofield (ed.), *Interpreting artefacts scatters: Contributions to plough zone archaeology*, 59-77. Oxford: Oxbow Books.

Gillings, M. & Sbonias, K. 1999. Regional survey and GIS: The Boeotia project. In M. Gillings, D.J. Mattingly & J. Van Dalen (eds.), *Geographical Information Systems and landscape archaeology*, 35-54. Oxford: Oxbow Books.

Given, M. 2004a. From density counts to ideational landscapes: Intensive survey, phenomenology and the Sydney Cyprus Survey Project. In E.F. Athanassopoulos & L. Wandsnider (eds.), *Mediterranean archaeological landscapes: Current issues*, 165-182. University of Pennsylvania, Museum of Archaeology and Anthropology.

Given, M. 2004b. Mapping and manuring: Can we compare sherd density figures? In S.E. Alcock & J.F. Cherry (eds.), *Side-by-side survey: Comparative regional studies in the Mediterranean world*, 13-21.Oxford: Oxbow Books.

Haselgrove, C., Millett, M. & Smith, I. (eds.) 1985. *Archaeology from the ploughsoil: Studies in the collection and interpretation of field survey data*. Sheffield: Equinox.

Hodder, I. & Orton, C.1976. *Spatial analysis in archaeology*. Cambridge: Cambridge University Press.

Kintigh, K.W. & Ammerman, A.J. 1982. Heuristic approaches to spatial analysis in archaeology. *American Antiquity* 47, 31-63. doi: 10.2307/280052

Marchi, M.L. 2010. *Ager Venusinus II (IGM 175 II SO; 187 I NO; 187 I SE; 188 IV NO; 188 IV SO)* (Forma Italiae 43). Firenze: Leo S. Olschki.

Marchi, M.L. 2016a. Carta Archeologica d'Italia – "Forma Italiae" project: Research method. In *LAC 2014 Proceedings (Rome 2014)*. doi: http://dx.doi.org/10.5463/lac.2014.42

Marchi, M.L. 2016b. Città, territorio e paesaggi antichi. Il contributo della Carta Archeologica per la ricostruzione delle trasformazioni ambientali. In G. Bruno & P. Carveni (eds.), *Atti del Convegno Nazionale Geositi, Geomorfositi e Geoarcheositi patrimonio geologico-ambientale del Mediterraneo (Portopalo di Capopassero 2015), Geologia dell'Ambiente* 3 suppl., 139-145. http://www.sigeaweb.it/documenti/gda-supplemento-3-2016.pdf

Marchi, M.L. & Forte, G. 2016. The GIS for the Forma Italiae project. From the GIS of the Ager Venusinus project to the GIS of the Ager Lucerinus project. Evolution of the system. In S. Campana, R. Scopigno, G. Carpentiero & M. Cirillo (eds.), *Keep the Revolution Going, Proceedings of the 43rd Annual CAA Conference (Siena 2015)*, 293-301. Oxford: Archaeopress.

Marchi, M.L. & Sabbatini, G. 1996. *Venusia (IGM 187 I NO / I NE)* (Forma Italiae 37). Firenze: Leo S. Olschki.

Millett, M. 1991. Pottery: Population or supply pattern? The ager Tarraconensis approach. In G. Barker & J. Lloyd (eds.), *Roman landscapes: Archaeological survey in the Mediterranean region*, 18-26. London: British School at Rome.

Nance, J.D. 1983. Regional sampling in archaeological survey: The statistical perspective. *Advances in Archaeological Method and Theory* 6, 289-356.

Orton, C. 2004. Point pattern analysis revisited. *Archeologia e Calcolatori* 15, 299-315.

Pelgrom, J. 2008. Settlement organization and land distribution in Latin colonies before the second Punic war. In L. De Ligt & S.J. Northwood (eds.), *People, land and politics. Demographic developments and the transformation of Roman Italy 300 B.C.-A.D. 14*, 317-356. Leiden: Brill.

Pelgrom, J. 2012. *Colonial landscapes. Demography, settlement organization and impact of colonies founded by Rome (4th-2nd centuries B.C.)*. PhD dissertation. Leiden University: Leiden University Repository.

Pelgrom, J. 2013. Population density in mid-Republican Latin colonies: A comparison between text-based population estimates and the results from survey archaeology. *Atlante Tematico di Topografia Antica* 23, 73-84.

Pelgrom, J., Marchi, M.L., Cantoro, G., Casarotto, A., Hamel, A., Lecce, L., García Sánchez, J. & Stek, T.D. 2014. New approaches to the study of village sites in the territory of Venosa in the Classical and Hellenistic periods. *Agri centuriati* 11, 31-60. doi: 10.1400/233577

Rathbone, D.W. 1981. The development of agriculture in the 'Ager Cosanus' during the Roman Republic: Problems of evidence and interpretation. *The Journal of Roman Studies* 71, 10-23. doi: 10.2307/299493

Rathbone, D.W. 2008. Poor peasants and silent sherds. In L. De Ligt & S.J. Northwood (eds.), *People, land and politics. Demographic developments and the transformation of Roman Italy, 300 B.C.-A.D. 14*, 305-332. Leiden: Brill.

Sabbatini, G. 2001. *Ager Venusinus I. Mezzana del Cantore (IGM 175 II SE)* (Forma Italiae 40). Firenze: Leo S. Olschki.

Salmon, E.T. 1969. *Roman Colonization under the Republic*. London: Thames and Hudson.

Schofield, A.J. (ed.) 1991. *Interpreting artefacts scatters. Contributions to ploughzone archaeology*. Oxford: Oxbow Monograph 4.

Settis, S. (ed.) 1984. *Misurare la terra: centuriazione e coloni nel mondo Romano*. Modena: Edizioni Panini.

Shennan, S., Gardiner, J. & Oake, M. 1985. *Experiments in the collection and analysis of archaeological survey data: The East Hampshire survey*. Sheffield: University of Sheffield.

Sommella, P. 2009. Esperienze documentali sul territorio dagli anni '80 ad oggi. Alcune considerazioni. *Archeologia e Calcolatori* 20, 47-59.

Stek, T.D. 2009. *Cult places and cultural change in Republican Italy. A contextual approach to religious aspects of rural society after the Roman conquest*. Amsterdam: Amsterdam University Press.

Stek, T.D., Modrall, E.B., Kalkers, R.A.A., Van Otterloo, R.H. & Sevink, J. 2015. An early Roman colonial landscape in the Apennine mountains: Landscape archaeological research in the territory of Aesernia (Central-Southern Italy). *Analysis Archaeologica: An International Journal of Western Mediterranean Archaeology* 1, 229-291.

Stek, T.D. & Pelgrom, J. 2013. Landscapes of early Roman colonization: Non-urban settlement organization and Roman expansion in the Roman Republic (4th-1st centuries B.C.). *Tijdschrift voor Mediterrane Archeologie* 50, 87.

Stek, T.D. & Pelgrom, J. (eds.) 2014. *Roman Republican colonization. New perspectives from archaeology and ancient history*, Papers of the Royal Netherlands Institute in Rome 62. Roma: Palombi.

Tarquini, S., Isola, I., Favalli, M., Mazzarini, F., Bisson, M., Pareschi, M.T. & Boschi, E. 2007. TINITALY/01: A new triangular irregular network of Italy. *Annals of Geophysics* 50 (3), 407-425. doi: 10.4401/ag-4424

Tarquini, S. & Nannipieri, L. 2017. The 10 m-resolution TINITALY DEM as a trans-disciplinary basis for the analysis of the Italian territory: Current trends and new perspectives. *Geomorphology* 281, 108-115. doi: 10.1016/j.geomorph.2016.12.022

Tarquini, S., Vinci, S., Favalli, M., Doumaz, F., Fornaciai, A. & Nannipieri, L. 2012. Release of a 10-m-resolution DEM for the Italian territory: Comparison with global-coverage DEMs and anaglyph-mode exploration via the web. *Computers & Geosciences* 38 (1), 168-170. doi: 10.1016/j.cageo.2011.04.018

Terrenato, N. 2000. The visibility of sites and the interpretation of field survey results: Towards an analysis of incomplete distributions. In R. Francovich, H. Patterson & G. Barker (eds.), *Extracting meaning from plough soil assemblages*, 60-71. Oxford: Oxbow Books.

Terrenato, N. 2007. The essential countryside: The Roman world. In S.E. Alcock & R. Osborne (eds.) *Classical Archaeology*, 139-161. Malden, MA: Blackwell.

Terrenato, N. & Ammerman, A.J. 1996. Visibility and site recovery in the Cecina valley survey, Italy. *Journal of Field Archaeology* 23 (1), 91-109. doi:10.1179/009346996791973990

Van de Velde, P. 1996. Off-site probleemstellingen, non-site technieken: De Riu Mannu survey op Sardinie. *Tijdschrift voor Mediterrane Archeologie* 17, 22-29.

Van De Velde, P. 2001. An extensive alternative to intensive survey: Point sampling in the Riu Mannu Survey Project, Sardinia. *Journal of Mediterranean Archaeology* 14 (1), 24-52. doi: 10.1558/jmea.v14i1.24

Van Dommelen, P. & Terrenato, N. 2007. Local cultures and the expanding Roman Republic. In P. Van Dommelen & N. Terrenato (eds.), *Articulating local cultures, power and identity under the expanding Roman Republic*, 7 – 12. Journal of Roman archaeology, JRA supplementary series number 63, Portsmouth.

Van Leusen, P.M. 1996. Unbiasing the Archaeological Record. *Archeologia e Calcolatori* 7, 129-135.

Van Leusen, P.M. 2002. *Pattern to process: Methodological investigations into the formation and interpretation of spatial patterns in archaeological landscapes*. PhD Dissertation, University of Groningen: University of Groningen Research Database.

Verhoeven, A.A.A. 1991. Visibility factors affecting artifact recovery in the Agro Pontino Survey. In A. Voorrips, S.H. Loving & H. Kamermans (eds.), *The Agro Pontino survey project: Methods and preliminary results*, 87-97. Amsterdam: Studies in Prae- en Protohistorie 6.

Waagen, J. 2014. Evaluating background noise: Assessing off-site data from field surveys around the Italic sanctuary of S. Giovanni in Galdo, Molise, Italy. *Journal of Field Archaeology* 39 (4), 417-429. doi: 10.1179/0093469014Z.00000000099

Wilson, A. 2008. Site recovery rates and the ancient population of the Biferno Valley. In G. Lock & A. Faustoferri (eds.), *Archaeology and landscape in Central Italy: papers in memory of John A. Lloyd*, 233-253. Oxford: School of Archaeology, University of Oxford.

Witcher, R. 2006. Broken pots and meaningless dots? Surveying the rural landscapes of Roman Italy. *Papers of the British School at Rome* 74, 39-72. doi: 10.1017/S0068246200003226

Witcher, R. 2011. Missing persons? Models of Mediterranean regional survey and ancient populations. In A. Bowman & A. Wilson (eds.), *Settlement, Urbanization, and Population*, 36-75. Oxford: Oxford University Press.

Chapter 3. Testing settlement models in the early Roman colonial landscapes of Venusia (291 B.C.), Cosa (273 B.C.) and Aesernia (263 B.C.)[14]

Abstract
This paper examines settlement density and settlement patterns in the Roman colonial territories of Venusia, Cosa and Aesernia, located in three different landscapes of central southern Italy (modern Basilicata, Tuscany and Molise). Using a series of GIS tools, we conducted a comparative analysis of the density and spatial distribution of sites dating to the Hellenistic period (*c.* 350–50 B.C.). We used the legacy settlement data collected by previous large-scale, intensive, site-oriented field surveys to test the validity of two competing rural settlement models of early Roman colonization: the conventional model of neatly organized settlements regularly dispersed across the landscape and the recently proposed theory that colonists adopted a polynuclear settlement strategy. After calculating the extent to which the archaeological datasets conform to the regular or polynuclear model, we concluded that only a very small portion of the colonized areas actually meets traditional expectations regarding the organization of early colonial settlements. Our analyses show that the legacy survey data are more consistent with the polynuclear settlement theory, but the data also reveal some completely unexpected patterns, suggesting that early Roman colonial landscapes were more diverse than previously thought.

3.1 INTRODUCTION

Roman colonization is traditionally depicted as an impressive enterprise that entailed the drastic reorganization of conquered territory (*cf.* Salmon 1969). In this view, Roman colonists lived in newly established towns that mimicked Rome (*e.g.* Brown 1980). The vast majority of colonists, however, would have settled in the hinterland of the colonial center, which is conventionally imagined by scholars to have been neatly partitioned (typically by centuriation) and characterized by a dense and regular distribution of colonists' farms. Over the last decades, as part of the broader development of landscape archaeology, archaeologists using field survey methods have intensively researched many areas of Italy that were affected by Roman colonization. On the basis of these large datasets, scholars have drawn important inferences about Roman settlement organization, the impact of Roman expansionism on conquered areas, the Roman economy and the relationship between these aspects (*e.g.* Goodchild & Witcher 2010; Launaro 2011; on legacy survey data see Witcher 2008).

Several salient problems emerge from the legacy datasets when we assess them against historical information about Roman colonization. In particular, if we compare the datasets against literary information about the number of colonists sent to the

---

14 This Chapter corresponds to the article "Testing settlement models in the early Roman colonial landscapes of Venusia (291 B.C.), Cosa (273 B.C.) and Aesernia (263 B.C.)" by Anita Casarotto, Jeremia Pelgrom & Tesse D. Stek, originally published (10 August 2016) in *Journal of Field Archaeology*, volume 41 issue 5 (publisher Taylor & Francis) as an Open Access article distributed under the terms of the Creative Commons Attribution-NonCommercial-NoDerivatives License (https://creativecommons.org/licenses/by-nc-nd/4.0/). This article is accessible through this link: https://doi.org/10.1080/00934690.2016.1211474 The legend of figures 3.2-3.4 has been adjusted to better explain the displayed information.
The authors of the article would like to thank Dr. Hans Kamermans, Arthur Hamel, Dr. Marleen Termeer, Dr. Jesús García Sánchez and Rogier Kalkers for their advice on our analyses and on the process of writing an early draft of the paper. Our gratitude goes also to Prof. Maria Luisa Marchi and Prof. Elizabeth Fentress who have made available for us seminal survey and ceramic data of the Ager Venusinus and Cosanus. Dr. Hans Kamermans has kindly taught the first author how to apply the Attwell-Fletcher test of association, which turned out very useful for one of the presented analyses and for her research in general.

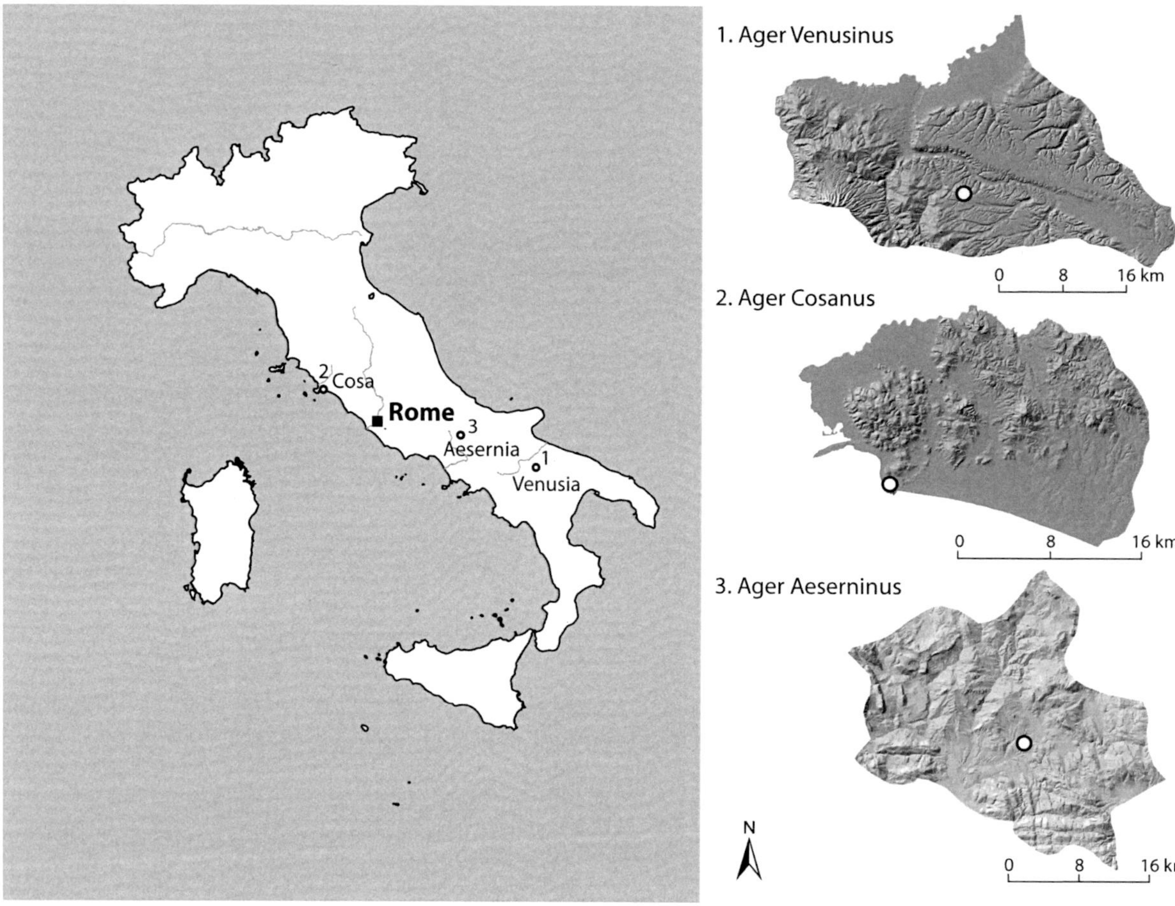

Fig. 3.1 Location of the three Latin colonies and the extent of their territories. The raster base map of the three territories (on the right) is the shaded relief calculated from the 10 m-resolution DEM named TINITALY/01 (Tarquini *et al.* 2007; 2012).

territories in question (Pelgrom 2008; 2012; 2013), it is evident that, as a rule, field surveys have mapped only a fraction of early colonial sites. In the past, these extremely low recovery rates were attributed to the methodological difficulties of recognizing small, simple rural dwellings in the survey record (Rathbone 1981; 2008; Millett 1991; Cambi 1999; Witcher 2011). Recently, however, an alternative solution to the "missing sites" problem has been suggested, namely, that colonial communities may have adopted settlement strategies that significantly differ from those conventionally envisaged. In a series of articles, two of the authors of this paper have critically reexamined the archaeological and epigraphic evidence of early Roman colonial settlement organization and have proposed an alternative polynuclear settlement scenario, in which colonists settled in large rural settlements, such as villages, separated by wide tracts of much more thinly populated land (Pelgrom 2008; 2014; Stek 2008, 166–215; 2009, 133–170; 2014). This article is part of a NWO-funded project (Netherlands Organization for Scientific Research) that aims to test the viability of this alternative hypothesis by combining a reassessment of the legacy data with new fieldwork (Stek & Pelgrom 2013).

The two opposing colonial settlement models, which have radically different spatial and social implications, are underpinned by the same datasets, namely those produced by previous regional surveys and published in the form of site distribution maps.

Accordingly, the key issue in this debate is the pattern that these large datasets actually present, not the quality of the datasets themselves. In this paper, therefore, we focus on density and pattern analysis using quantitative statistical GIS tools to establish which model of settlement organization the survey data truly support, while suggesting other settlement models for consideration along the way. We include in our analysis the data collected in the territory of three intensively studied Latin colonies: Venusia (founded in 291 B.C.), Cosa (273 B.C.) and Aesernia (263 B.C.) (Fig. 3.1). In light of the centrality of these datasets for other important debates on Roman society, such as the nature of the Roman economy, town-countryside relations, demography and the nature of Roman imperialism, the significance of the conclusions of this paper extends far beyond the debate over Roman colonization.

3.2 DATA

The present analysis capitalizes on the rich datasets compiled during three regional field surveys carried out in the territory of the colonies of Venusia, Cosa and Aesernia (respectively published in Marchi & Sabbatini 1996; Sabbatini 2001; Marchi 2010; Carandini *et al.* 2002; Stek *et al.* 2015). These projects were executed in the late 1970s to mid-1980s (Cosa), in the late 1980s to mid-2000s (Venusia) and, more recently, from 2011 to the present (Aesernia).

A similar survey methodology, which may be described as large-scale, intensive and site-oriented, was used to collect the datasets. Teams composed of 3 to 5 surveyors spaced 5 to 10 m (Venusia and Aesernia projects) or 10 to 20 m (Cosa project) apart systematically walked through all accessible field units in the sample survey area. All observable scatters of archaeological material (site density set at $\geq$ 5 shards per sq m for the Venusia and Aesernia projects and a density scale of 1 to 5 for the Cosa project) were recorded on IGM maps (1:25,000), CTR maps (1:5000; 1:10,000) or by GPS. Concentrations of material were dated on the basis of diagnostic ceramics, samples of which were collected for laboratory analysis.

Despite the richness and high quality of these datasets, comparable to most reconnaissance research in other areas of the Mediterranean world (*e.g.* the southwest Argolid project [Jameson *et al.* 1994] and the Boeotia project [Bintliff & Snodgrass 1985]), there are obvious, much-discussed methodological problems with regional site-orientated surveys and with the validity of survey data in general, especially with respect to their completeness (*e.g.* Barker & Lloyd 1991; Terrenato & Ammerman 1996; Bintliff & Sbonias 1999; Fentress 2000; Terrenato 2000; Van Leusen 2002; Van Leusen *et al.* 2011).

Our focus is on critically analyzing the patterns that can be discerned in these legacy datasets. This is relevant not only because these datasets have significantly influenced several important past and current historical and archaeological discourses, but also because different conclusions have been drawn from them, suggesting radically different settlement scenarios. As we will demonstrate, contrary to what scholars have suggested in the past, these datasets do not corroborate the conventional model of Roman colonial settlement organization, but are reasonably consistent with other types of settlement organization.

Our analyses include all settlement sites recorded inside the proposed colonial territories (Table 3.1) broadly datable to the Hellenistic period (*c.* 350–50 B.C.), primarily on the basis of the presence of black-gloss pottery. We adopt this rough chronological range to study early colonial settlement patterns for both practical and theoretical reasons. First, the number of sites that can be dated precisely to the early colonial phase (*i.e.* 3rd century B.C.) is too small to identify statistically significant patterns. This is a well-known problem that scholars have typically addressed by including a category of possible sites (defined over a broad chronology, *e.g.* sites generically defined as Hellenistic or Republican settlements) to compensate for the underrepresentation of poorly detectable site types or periods (for this approach, see Goodchild & Witcher 2010, 196–198). Including potentially later sites may distort our understanding of early settlement patterns, since 2nd- and 1st-century developments may have differed from earlier conditions. However, since the inclusion of potentially later sites is likely

Table 3.1 Legacy data and samples.

Colony	Modern Province and Region	Survey project	Project survey area (sq km)	Area of the proposed colonial territory (sq km)	Colonial territory inside survey area (sq km)	Early colonial settlements	Hellenistic settlements[1]
Venusia	Potenza (Basilicata)	Forma Italiae (Marchi & Sabbatini 1996; Sabbatini 2001; Marchi 2010)	700	890 (based on Coppa 1979)	530	78 0.1 sites per sq km	564 1.1 sites per sq km
Cosa	Grosseto (Tuscany)	Paesaggi d'Etruria (Carandini *et al.* 2002)	300	430 (based on Cardarelli 1924 -1925)	135	73 0.5 sites per sq km	200[2] 1.5 sites per sq km
Aesernia	Isernia (Molise)	Landscapes of Early Roman Colonization (LERC) (Stek *et al.* 2015)	120	680 (based on Toynbee 1965)	120	4 0.03 sites per sq km	81 0.7 sites per sq km

[1] In all three case studies, certain, probable and possible Hellenistic settlements are taken into account.

[2] On occasion, when two or more sites were found very close to each other, the Cosanus survey team assigned them the same UTM coordinates (reported in Carandini *et al.* 2002, 379–409). Since we used these coordinates for the site digitalization, sites with the same position appear as a single dot, both in the original distribution maps (published in Carandini *et al.* 2002) and in the Figures 3.3, 3.6, and 3.10B. In the following density and pattern analyses, however, they are counted as distinct sites.

to strengthen rather than weaken the conventional scenario (with higher site densities and more regular patterning), any indication of divergent patterns in the aggregate data (such as clustering) takes on even greater significance. Again, we emphasize that the central aim of this paper is to investigate both the potential of the unfiltered legacy survey data available and the robustness of the various settlement models that have been inferred from them. As a matter of fact, we believe that further detailed research on colonial sites and finds (see Pelgrom *et al.* 2014; Stek *et al.* 2015), and on possible biasing factors related to field-survey recording methods (such as ground visibility and geomorphological processes), is needed to confirm the validity of the patterns identified in this paper (see Chapters 4 and 5).

3.3 METHODS

We systematically analyzed the existing survey data on two interrelated levels: the density and spatial configuration of settlement sites. To avoid potential bias from previous scholars' categorization of sites (*i.e.* function or size), all sites are visualized in GIS as simple, unclassified dots: that is, Hellenistic settlements are represented only by their centroids. On account of the long occupational history of a majority of these sites, their extension (which was recorded during field survey) is not necessarily indicative of the early colonial phase. Therefore, it would be incorrect to use the information about the size of these settlements so as to distinguish potential colonial farms from larger settlement types (such as villages or villas). Moreover, the scatter size is also strictly depending on invasive plowing activities which can transform small, dense concentrations of archaeological finds into larger, more diffuse scatters (*e.g.* Shennan *et al.* 1985; Fentress 2000; Van Leusen 2002; Given 2004; Feiken 2014). Therefore, unclassified and homogeneous point distributions are considered the primary evidence for quantitatively testing associated site density and patterns (Orton 2004). It is important to stress that these methodological decisions favor the conventional colonial settlement model, since the pluriform archaeological reality is reduced to equally sized dots in conformance with the notion of regularly settled landscapes dotted with mononuclear farmsteads.

The spatial analyses were conducted in two steps. First, the conventional model, which expects high density of sites in the colonial countryside, was tested in ArcGIS (version 10.2.2) by means of a point-density analysis, which calculates the number of sites per square kilometer (Esri 2014a). In order to broaden the scope beyond the rival theories of dispersed or nucleated colonial settlements, we also considered alternative scenarios that might explain the density patterns recorded. In particular, we analyzed whether the primary urban settlement of the colony influenced rural settlement densities as predicted by Von Thünen's Isolated State model. The Von Thünen model (1966 [1826]) predicts a gradual decline in settlement density as the travel cost to reach the urban center increases, which is correlated with different agricultural practices utilized in concentric land-use bands around the urban center. This model has already been applied to better understand whether settlement density correlates with distance to city; for example, in Patterson's study (2004) of the Roman economy in the Tiber valley (see also De Neeve 1984, 10–16 and Morley 1996, 11, 58–82 for a discussion in economic terms of the Von Thünen model in Roman contexts). Patterson demonstrates that, despite its abstract and reductive nature, the model closely corresponds to the data on Early Imperial Rome and its hinterland (but see Horden & Purcell 2000, 112–122 and Witcher 2009, 477–478 for a critical position): higher site density was identified closer to Rome, which matches Von Thünen's theory of the intense exploitation and settlement of rural areas closer to urban markets. This was also the main expectation we tested with our datasets.

In a second step, we conducted a point-pattern analysis to detect regular, random or aggregated patterns (Hodder & Orton 1976, 30–98; Kintigh & Ammerman 1982; Roberts 1996, 56–57). Translating the two competing colonial settlement models into spatial-analytical terms, we then tested for the presence of either a regular (conventional model) or clustered (polynuclear model) settlement pattern. In this analysis, we focused on the interactions between settlements over local distances (so-called second-order effects in pattern analysis [*cf.* Orton 2004; Bevan & Conolly 2006; Palmisano 2013]). In essence, by looking at how settlements are located in relation to other settlements (and over what distances), we aimed to identify dispersed (regular distributions of farms) or agglomerative (villages) processes underpinning the colonial settlement system in different parts of the landscape. The influence of environmental and cultural landscape characteristics on settlement location preferences will be analyzed in Chapter 6 (*i.e.* first-order effects: on how to incorporate them in pattern analysis see Bevan & Conolly 2006, 229–230; Winter-Livneh *et al.* 2010, 288–293; Bevan & Wilson 2013; Palmisano 2013; see also how predictive modeling investigates first-order effects in Judge & Sebastian 1988; Kvamme 1990; Van Leusen & Kamermans 2005; Verhagen 2007).

We applied a Multi-Distance Spatial Cluster analysis (global Ripley's K-function) in ArcGIS to highlight statistically significant clustered or dispersed patterns over a wide range of scales of analysis (Esri 2014b). This statistical tool graphically illustrates how the spatial arrangement of dots changes as the scale changes (thus in tandem with the size of the study area being evaluated around the dots). Although the more popular Nearest Neighbor analysis by Clark & Evans (1954) also is a valuable approach to point-pattern analysis (see Chapter 2), we feel that it is less effective in evaluating the patterns in our datasets. The Nearest Neighbor analysis, indeed, calculates a clustering index based only on the mean Euclidean distance from each site to the next nearest (Bailey & Gatrell 1995, 90–91), whereas the global Ripley's K-function measures the average number of sites from each source site within several given distances (*i.e.* at every scale of analysis); it thus can identify significant correlations (if any) between distribution and density at various scales of analysis. On the contrary, since the Nearest Neighbor analysis considers only one scale corresponding either to the extent of the sample area (in our case, the survey sample area of the colonial territory) or to a rectangle enclosing all the dots, it can detect only the dominant pattern characterizing the entire study area sample.

To answer the question posed in this paper, it is crucial to deal with the fact that the pattern changes as the scale used to observe the spatial distribution changes (see the discussion in Lock & Molyneaux 2006). As way of example, a cluster of dots will appear as composed by dispersed points if we look at it at a narrow scale. In reverse, as we scale up, the cluster configuration will again become apparent. The Ripley's K-function calculation offers a formal approach to this problem, which is why it is best suited to our analysis: it takes into account several scales of analysis (*i.e.* progressively larger sample areas around the dots) and thus has control over pattern variability as the scale changes.

The Ripley's K-function is calculated as $K(d)= E/\lambda$, where E is the number of events within distance d from a randomly chosen centroid location, and λ is the average intensity of events per unit area (Ripley 1976; Bailey & Gatrell 1995, 92–94; Dixon 2002, 1796; Bevan & Conolly 2006, 220–221; Palmisano 2013, 350; Sayer & Wienhold 2013, 77). In practice, this tool evaluates the dot pattern at several progressively greater distances from the source sites: it computes the degree of clustering or dispersion by comparing the average number of neighboring sites from each source site, at every distance being evaluated (scale of analysis), with the average density of sites throughout the study area (Esri 2014b).

As part of this pattern analysis we then selected physically and culturally uniform areas with constantly high point density that are large enough to support farm-based intensive agriculture. The goal then was to test, on a small-scale and with environmental and cultural conditions as equal as possible, whether an even distribution of farms manifests itself in the data. This is, indeed, the typical settlement organization pattern anticipated in the conventional colonial model: in principle, wherever favorable physical conditions (*i.e.* smooth/flat morphology) in the conquered territory allow for orderly land partition, we should expect to find farms distributed regularly across the terrain, corresponding to a highly precise, grid-based allotment system and, thus, dependent exclusively on regular local distances between farms (*i.e.* second-order effects).

As a last step of this pattern analysis, we also carefully examined the cluster pattern that seems to characterize large tracts of colonial territories. We tested further by using Bintliff's approach to the study of territorial organization (Bintliff 1999; 2000; 2009). First, we indicated possible early colonial nucleated settlements (or villages) and then analyzed the polynuclear configuration of the dot clustering. Bintliff's socio-ecological approach to modeling Mediterranean village-based systems serves as a useful *comparandum* as we calibrated the pattern of our datasets and established parallels (if any) to a nucleated form of settlement (Roberts 1996, 15–37).

In contrast to Bintliff's studies, our analysis does not focus on the definition of territorial catchments (Vita-Finzi & Higgs 1970), but rather operates on a simpler descriptive level whereby distances between colonial settlement clusters are compared with Bintliff's distance predictions. He identifies settlement catchment radii of comparable length according to different degrees of rural infill (and demographic growth) and different fission levels of the village system at issue (related to dynamics of territorial competition). In the Ager Venusinus and the Ager Cosanus, we analyzed the distance between neighboring hotspots with high localized settlement density (potential villages) to see how these distances match the standard inter-distances proposed by Bintliff.

3.4 TESTING SETTLEMENT DENSITY: POINT-DENSITY ANALYSIS

According to the conventional understanding of Roman colonization, a majority of the colonists sent by Rome to populate freshly conquered lands settled in farms on individual plots carved out of the *ager* of the colony (*i.e.* the territory under the jurisdiction of the colony). Colonial urban centers were small and thus could have hosted only a limited number of colonists. Literary evidence on colonial populations

Table 3.2 Percentages of rural territory defined by different settlement density.

Density (sites per sq km)	Ager Venusinus		Ager Cosanus		Ager Aeserninus	
	With unfeasible survey conditions	Without unfeasible survey conditions	With unfeasible survey conditions	Without unfeasible survey conditions	With unfeasible survey conditions	Without unfeasible survey conditions
d≥ 20	0.03 %	0.03 %	0	0	0	0
d≥ 10	0.79 %	0.84 %	0.385 %	0.40 %	0	0
d≥ 8	1.38 %	1.24 %	0.93 %	1.04 %	0.05 %	0.285 %
d≥ 5	4.81 %	4.29 %	4.44 %	4.15 %	0.78 %	2.55 %
d≥ 3	13.21 %	12.12 %	18.53 %	20.07 %	6.08 %	10.175 %
d≥ 1	45.535 %	43.56 %	55.31 %	59.86 %	36.58 %	54.41 %
d = 0	54.465 %	56.44 %	44.69 %	40.14 %	63.42 %	45.59 %
Total area (sq m)	530,066,147.317	396,770,118.377	135,337,975.878	81,257,583.851	120,401,979.430	14,977,976.518

and the size of the allotments they received, in particular Livy's *Ab Urbe condita*, suggests large colonial populations, usually ranging between 2500 and 6000 colonists. This translates into farm densities of at least eight farms per sq km, if one accepts a sensible urbanization percentage of 20–30% (for in-depth discussion on these estimates, see Garnsey 1979; Pelgrom 2008; 2013, 74–75; on estimating population densities for colonial landscapes, see Fentress 2009).

In order to test whether such densely populated landscapes are visible in the survey data and, if so, where they are located, we used the Point Density tool in ArcGIS. This tool estimates the density of points around each output raster cell in a user-defined neighborhood (Esri 2014a). In this case, we chose a circle of one sq km. This results in a raster surface in which the value of each cell (set at 20 by 20 m) represents the number of sites found in the circle. Cells with a density higher than twenty, ten, eight, five, three and one settlements per square km were then isolated and the extent of their respective areas was calculated. In this way, we calculated the percentage of the landscape that corresponds to the expected farm density of eight sites per square km.

The results shown in Table 3.2 clearly demonstrate that the extent of rural landscape characterized by a settlement density equal to or higher than eight sites per square km barely approaches 1% in all three case studies. Significantly, this enormous divergence from conventional expectations even appears when we consider the "best-case" scenario for early colonial occupation. Even if we assume *all* broadly dated Hellenistic settlements are early colonial farms, the actual field survey data exhibit no observable correlations with historically expected site densities.

One explanation for this enormous discrepancy may be sought in adverse survey conditions, which may have prevented archaeologists from recording a majority of sites. Such factors as poor visibility of the walking surface or the frequent erosion of the steepest slopes may have had a detrimental effect on site detection (these factors are discussed at length in Chapters 4 and 5). As a result, the representativeness of the samples under consideration here may be significantly undermined (but see Chapters 4 and 5).

To assess the potential impact of such factors, we conducted a second point density analysis, this time excluding possible biased samples. As an experiment, we only considered zones and settlements located in modern arable land characterized by good surface visibility and gentle slope conditions, which is widely known to afford ideal surface visibility for site discovery. For the Ager Venusinus and Cosanus, the areas covered by forest, artificial surfaces and water bodies (extracted from the Corine Land Cover 2000 and 1990–1:100,000, Sambucini *et al.* 2010, 9–15), and slopes steeper than 20% (*cf.* FAO 2006, 11–12; Arnoldus-Huyzendveld 2007), were excluded from the colonial territory sample. We followed a broadly

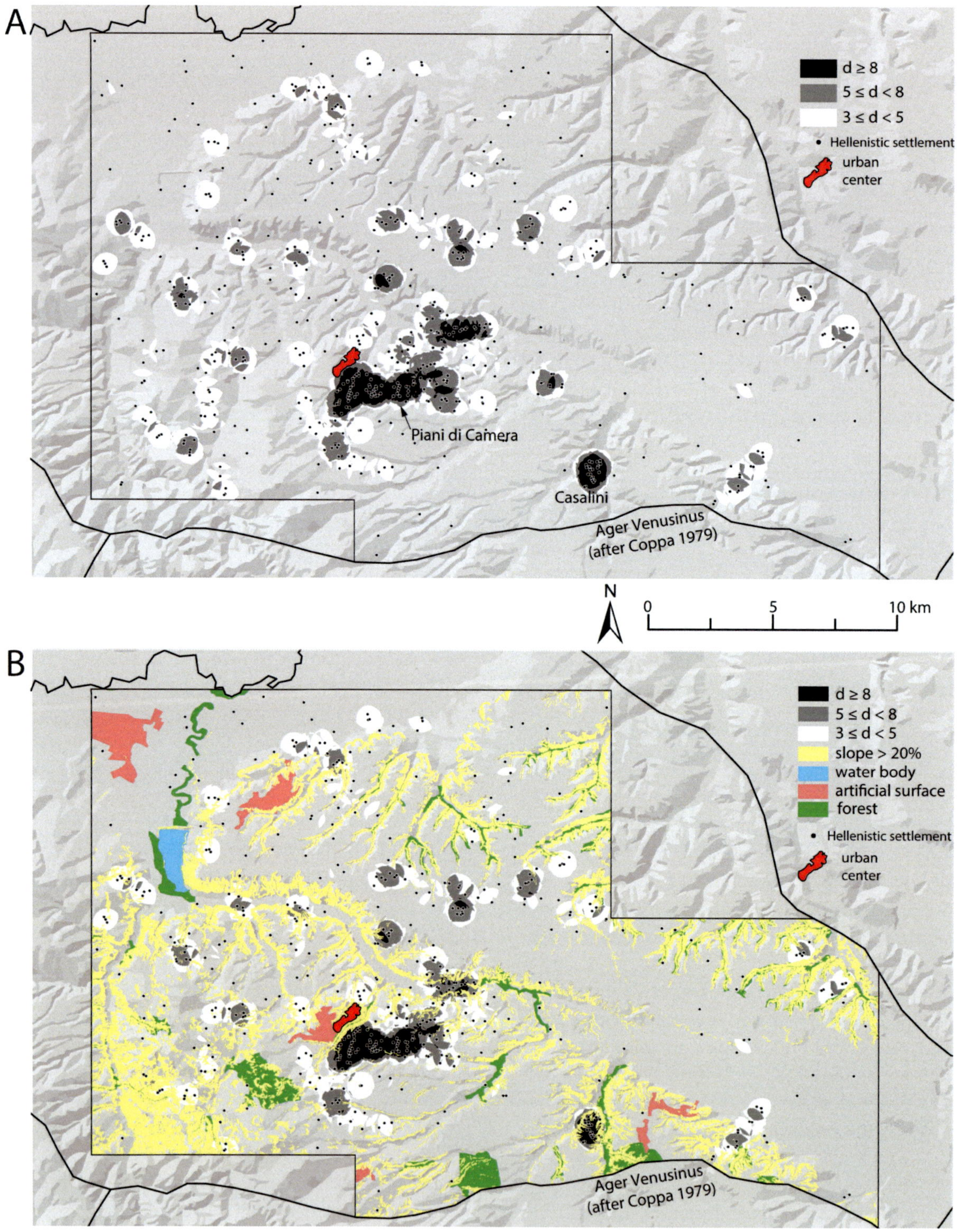

Fig. 3.2 A) Point-density analysis of the Hellenistic settlements (black dots) in the survey sample area of the Ager Venusinus; B) Point-density analysis excluding sites and zones in unfeasible survey conditions (forest, artificial surfaces, water bodies and slope > 20%). Base map: hillshade elaboration of the 10 m-resolution DEM named TINITALY/01 (Tarquini *et al.* 2007; 2012).

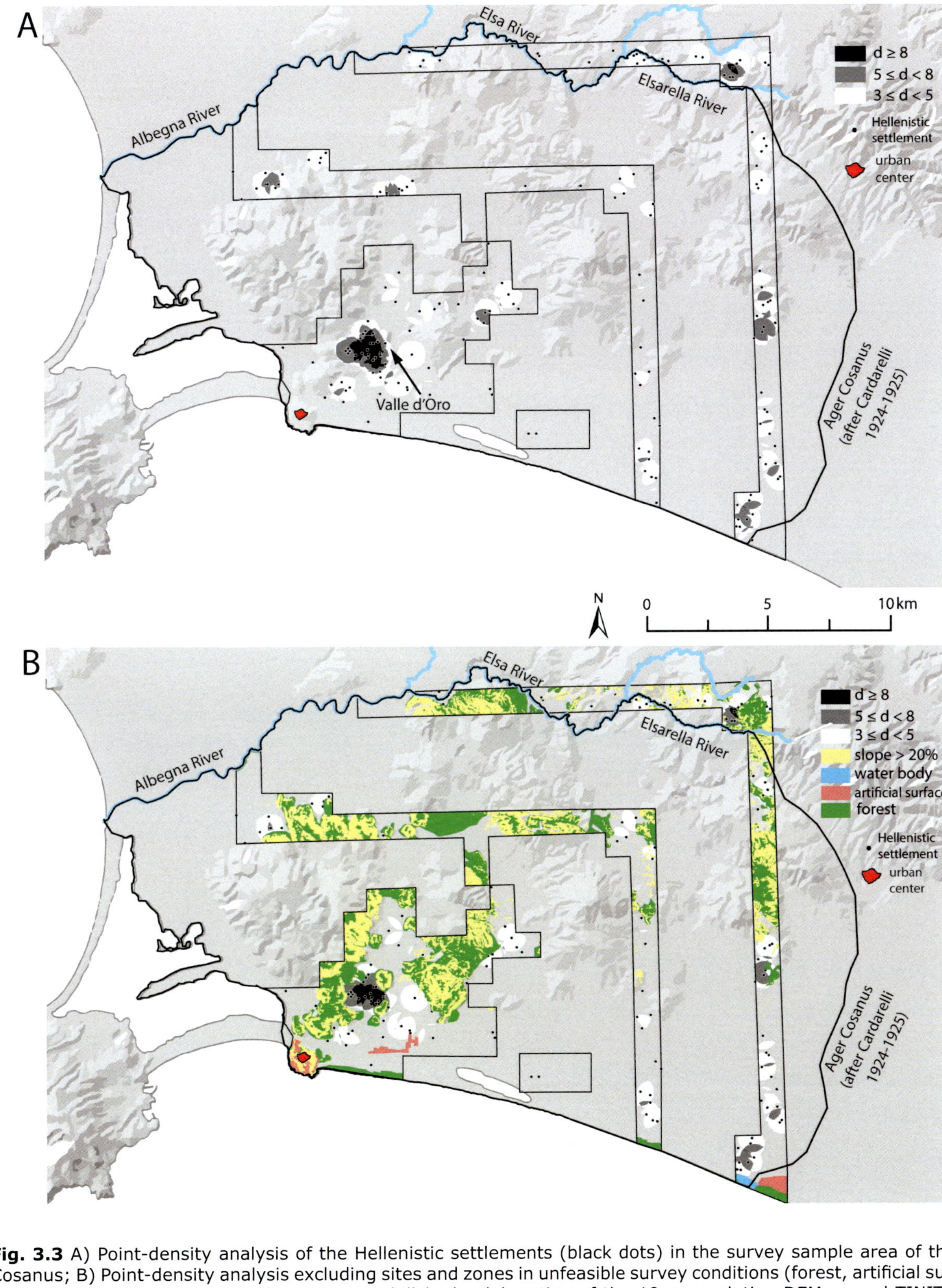

Fig. 3.3 A) Point-density analysis of the Hellenistic settlements (black dots) in the survey sample area of the Ager Cosanus; B) Point-density analysis excluding sites and zones in unfeasible survey conditions (forest, artificial surfaces, water bodies and slope > 20%). Base map: hillshade elaboration of the 10 m-resolution DEM named TINITALY/01 (Tarquini *et al.* 2007; 2012).

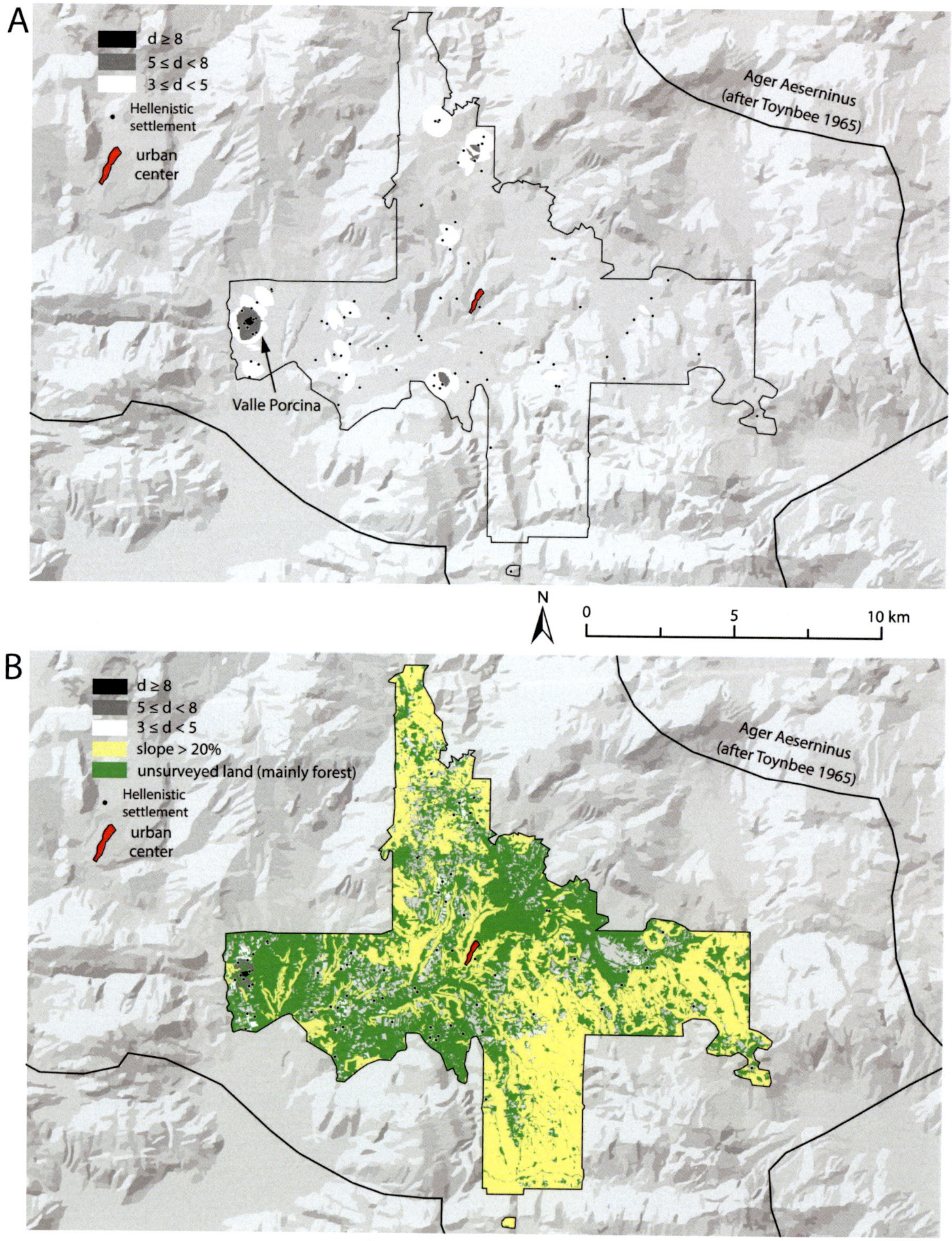

Fig. 3.4 A) Point-density analysis of the Hellenistic settlements (black dots) in the survey sample area of the Ager Aeserninus; B) Point-density analysis excluding sites and zones in unfeasible survey conditions (unsurveyed land and slope > 20%). Base map: hillshade elaboration of the 10 m-resolution DEM named TINITALY/01 (Tarquini *et al.* 2007; 2012).

similar but more precise procedure for the Ager Aeserninus, for which we have detailed information on the survey visibility conditions and land use of each unit walked. The number of Hellenistic sites located in the remaining zones with favorable field survey conditions in the colonial territories is respectively 493 in the Ager Venusinus, 164 in the Ager Cosanus and 69 in the Ager Aeserninus.

Despite our effort to exclude the most common adverse conditions for field surveys, the percentage of territory characterized by a site density equal to or higher than eight per square km remains very small (lower than 1.5%) in all the three colonial territories (Table 3.2). Even when we analyzed select samples of the field survey area and possibly more representative site samples, the traditional scenario of a radically reorganized, evenly dotted Roman countryside is virtually invisible in the archaeological record.

This major discrepancy between expected site densities and the survey record can readily be appreciated in Figs. 3.2–3.4. Areas with a density of eight or higher are very limited. Certain spatial patterns, however, are visible in the data. For example, site densities of five and higher are located primarily in fertile plains close to urban centers (*e.g.* the Piani di Camera in the case of Venusia, and the middle of the Valle d'Oro in the case of Cosa) and are scattered more widely the further away they are from the centers. In the Ager Venusinus, there is also an area in between these two "bands" of higher density that is relatively devoid of settlements. The spatial configuration of high-density areas in the territory of Aesernia is rather different (Fig. 3.4). The highest and most homogeneous site density is not located near the urban center, as in Venusia and Cosa, but rather is concentrated in a river valley (the Valle Porcina), far west of Aesernia.

3.4.1 TESTING TRENDS IN DENSITY: VON THÜNEN'S ISOLATED STATE MODEL

Cultural attractors, such as an urban center with its political and economic facilities, can influence land-use strategies and settlement density in the surrounding territory, varying according to the distance from them. In *Isolated State* (1966 [1826]), the German agronomist Von Thünen depicts an idealized scenario in which a market located in the middle of a flat isotropic landscape naturally tends to organize the surrounding hinterland in several concentric land use bands. Von Thünen proposes the following system of land use, moving from the town outwards: intensive production: horticulture and dairy-farming; silviculture; extensive agriculture (intensive arable rotation, arable with long ley, three-field arable); and ranching (Grotewold 1959; Chisholm 1968, 20–32; Haggett *et al.* 1977, 205–207; Goodchild 2007, 31–35). According to this model, settlement density decreases as the distance from the town increases.

We tested the Von Thünen density trend against the survey data. Since several variables, such as the topography, routes, rivers and secondary markets (Haggett *et al.* 1977, 211–222), may distort the Von Thünen's idealized land-ring pattern we adjusted the model accordingly (*cf.* Dodson 1991; Thornton & Jones 1998). In IDRISI GIS (Selva edition), we incorporated the effects of landscape morphology, arguably the most important factor on the movement of people, by implementing a cost analysis (using the VARCOST module; see Eastman 2012, 277–281) based on slope and aspect values, which are extracted from the 10 m–resolution DEM named TINITALY/01 (Tarquini *et al.* 2007; 2012). In conducting this cost analysis, we modeled the cost of moving from the city to its hinterland as a good approximation of the cost necessary to walk the other direction, from the hinterland to the city. Moreover, we treated distance simply as the physical distance people walked and did not consider the effect of moving different types of goods on transport costs (economic distance: see Zipf 1949; Chisholm 1968, 30).

After creating cost surfaces based on slope and aspect conditions (see also Wheatley & Gillings 2002, 151–159; Conolly & Lake 2006, 215–225), we divided the colonial territories into concentric land-use cost-bands centered around the city. To do this, we have accepted previous scholars' reconstructions of the colonial *agri* (Cardarelli 1924–1925; Toynbee 1965; Coppa 1979) as the maximum territorial

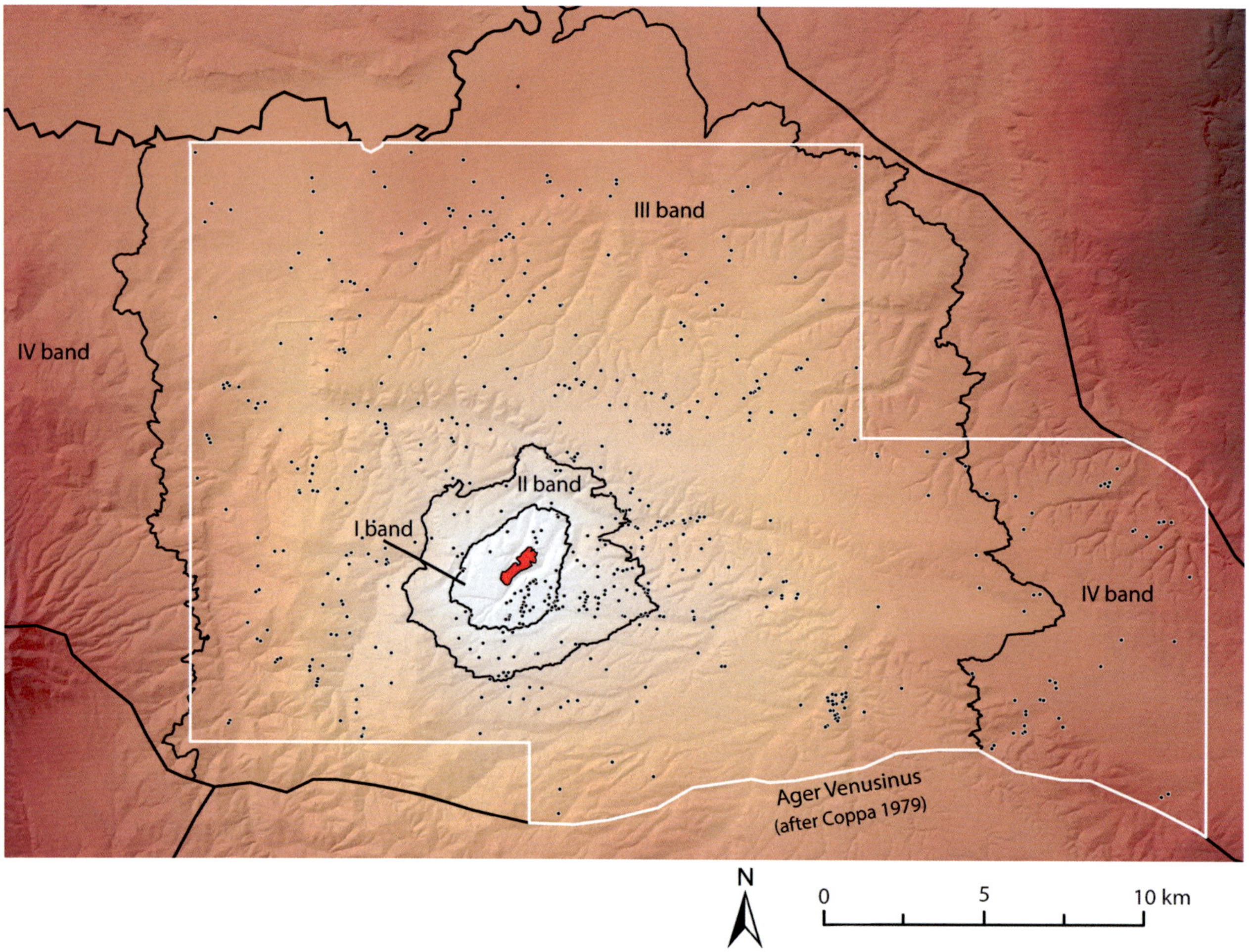

Fig. 3.5 Von Thünen's model implemented in the Ager Venusinus. Cost-surface created from the 10 m-resolution DEM named TINITALY/01 (Tarquini *et al.* 2007; 2012) and calculated in IDRISI GIS (VARCOST module): the increasing cost from the city to the hinterland ranges from low (white) to high (dark red).

extents of these colonies (for a critical discussion of these modern territorial reconstructions see Pelgrom 2014; Stek 2014). Within these territories, we distinguished the four main zones of agricultural activity as described above. If we apply the calculations of Haggett and colleagues (1977, 205), the first band (intensive agriculture) covers 1% of the territory, the second band (forest) 3%, the third band (extensive agriculture) 58% and the fourth band (ranching/grazing) 38%. We reclassified the cost surfaces accordingly. As a result, the *agri* are carved up into four land-use cost-bands (Figs. 3.5–3.7).

In a final step, we used the Attwell-Fletcher test of association (Attwell & Fletcher 1985; 1987; Kamermans 2000) to analyze the number of sites located in the cost-bands in the three survey sample areas: first, the number of settlements located in each band was compared to the percentage of surface surveyed in that band (*i.e.* the number of observed settlements was confronted with the proportion of settlements expected in that surface); second, significant associations (if any) were then indicated. The Attwell-Fletcher test evaluates whether the concentration of sites in each cost-band is positively significant (*i.e.* there are significantly more sites than expected from a random distribution:

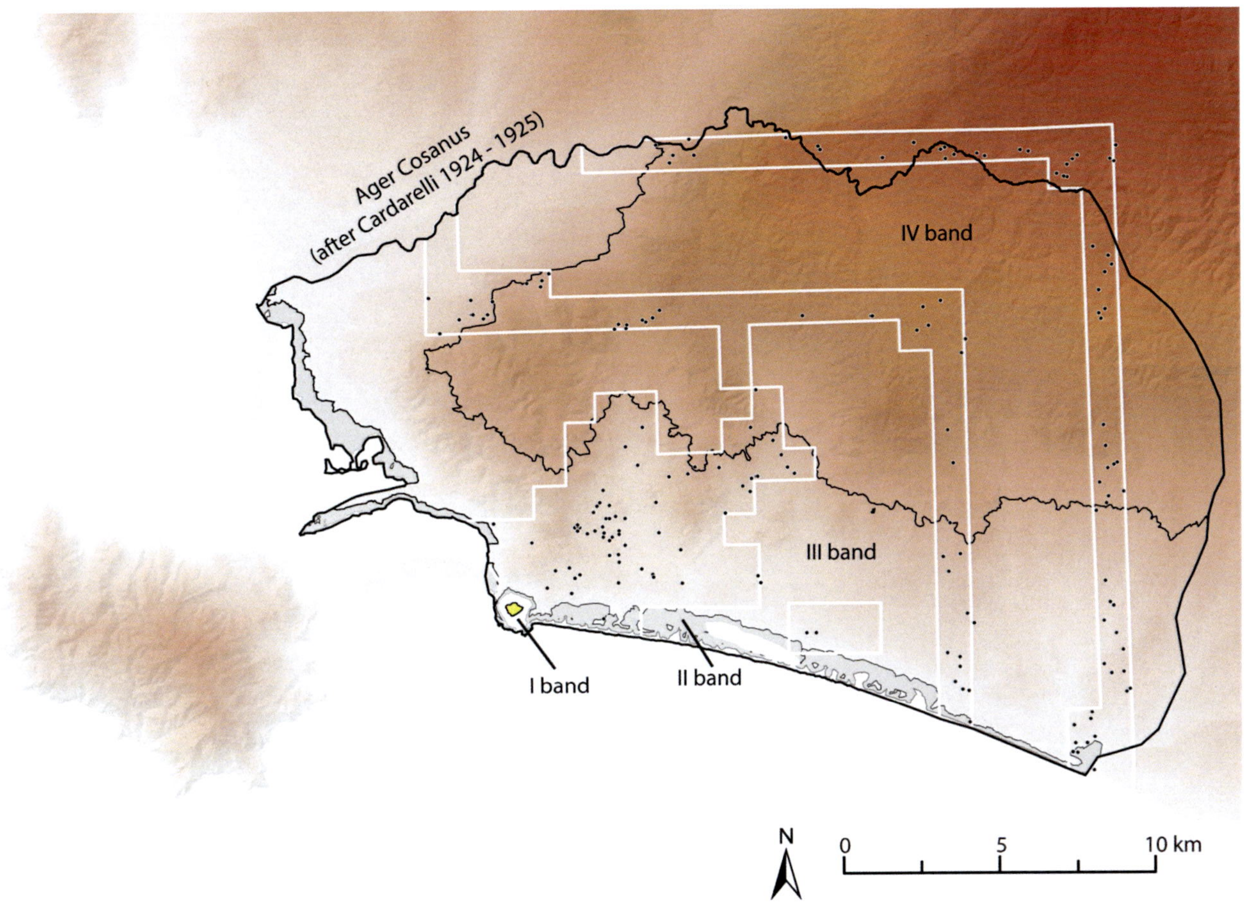

Fig. 3.6 Von Thünen's model implemented in the Ager Cosanus. Cost-surface created from the 10 m-resolution DEM named TINITALY/01 (Tarquini *et al.* 2007; 2012) and calculated in IDRISI GIS (VARCOST module): the increasing cost from the city to the hinterland ranges from low (white) to high (dark red).

category weight > than the critical value for 95th percentile), negatively significant (*i.e.* significantly fewer sites than expected: category weight < than the critical value for 5th percentile) or merely due to chance.

This statistical analysis permits us to recognize significant density patterns as cost-distances from the colonial town increase. As displayed in Tables 3.3 and 3.4, there is a significant tendency in the Ager Venusinus for settlements to cluster in the first concentric cost-band around the colonial town. Moreover, significant evidence of avoidance allows us to confidently infer that site concentration decreases significantly in the third and fourth zones. In the Ager Cosanus and Aeserninus, no significant correlations are found. This may be due to the smaller size and narrower and more irregular shape of the survey transects that do not allow for the observation of the pattern in extension (transects of *c.* 1 km wide regularly spaced with a wider one covering the Valle d'Oro are present in Cosa, a cross-shaped transect in Aesernia). Sample choices may also affect the following point-pattern analysis.

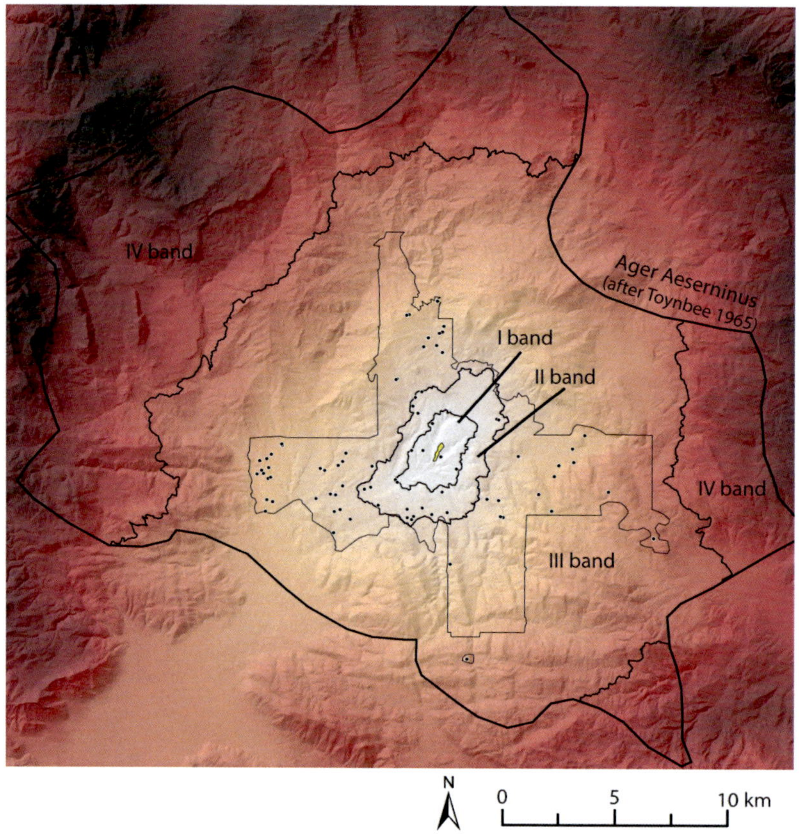

Fig. 3.7 Von Thünen's model implemented in the Ager Aeserninus. Cost-surface created from the 10 m-resolution DEM named TINITALY/01 (Tarquini *et al.* 2007; 2012) and calculated in IDRISI GIS (VARCOST module): the increasing cost from the city to the hinterland ranges from low (white) to high (dark red).

3.5 TESTING SETTLEMENT DISTRIBUTION: POINT-PATTERN ANALYSIS

We turn now from density to spatial configuration. Our aim here is to discern significant settlement patterns and to assess to which colonial scenario they correspond more closely. A regular pattern would nicely fit the conventional model, because it predicts farms located in regularly distributed modular plots (see Hudson 1969; Hodder & Orton 1976, 54–85; Perles 2001, 132–147 for a discussion of regular patterns in rural areas). A random (but dense) arrangement might also conform to the conventional scenario: in that case, random distribution may be explained by the variable position of farms within their allotments (Celuzza 1984, 159). A clustered pattern separated by tracts of empty space, however, might indicate a polynuclear, village-based settlement system.

In order to detect potential clustered or regularly dispersed site distribution patterns, a global K-function calculation was run in ArcGIS. We performed a Multi-Distance Spatial Cluster analysis with ArcGIS inside the survey sample areas of the colonial territories (thus excluding the unsurveyed zones outside the transects) and we applied the correction method to simulate outer boundary sites in order to limit the edge-effect problem (*i.e.* likely underestimation of sites next to the borders between surveyed and unsurveyed areas). The site densities found in increments of 50 m to 5000 m from each site were averaged and a Monte Carlo simulation consisting of 99 permutations of randomly distributed points was performed to set the confidence interval at 99% ($\alpha = 0.01$) (Winter-Livneh

Table 3.3 Outcomes of the Attwell-Fletcher test, comparing the number of settlements observed in each cost-band of the Venusian survey sample. Number of sites = 543 (the cluster of dots—22 points—in the 3rd band, representing the nucleated village of Casalini, is counted as a single point); number of categories = 4; number of simulations = 200. 95th percentile for max weight = 0.35 ± 0.016 ; 5th percentile for min weight = 0.14 ± 0.004.

Ager Venusinus (including sites and zones in unfeasible survey conditions)						
Cost band	**Number of sites**	**Expected proportion**	**Observed proportion**	**Category weight**	**More sites than expected**	**Fewer sites than expected**
1st	47	0.019	0.09	0.52	yes	no
2nd	81	0.052	0.15	0.32	no	no
3rd	375	0.792	0.69	0.10	no	yes
4th	40	0.137	0.07	0.06	no	yes

Table 3.4 Outcomes of the Attwell-Fletcher test, comparing the number of settlements observed in each cost-band of the Venusian survey sample. Number of sites = 471 (the cluster of dots—22 points—in the 3rd band, representing the nucleated village of Casalini, is counted as a single point; sites located in unsuitable survey conditions are excluded, in total 72); number of categories = 4; number of simulations = 200. 95th percentile for max weight = 0.37 ± 0.006; 5th percentile for min weight = 0.11 ± 0.021.

Ager Venusinus (excluding sites and zones in unfeasible survey conditions)						
Cost band	**Number of sites**	**Expected proportion**	**Observed proportion**	**Category weight**	**More sites than expected**	**Fewer sites than expected**
1st	46	0.015	0.10	0.61	yes	no
2nd	75	0.057	0.16	0.26	no	no
3rd	314	0.786	0.67	0.08	no	yes
4th	36	0.142	0.08	0.05	no	yes

et al. 2010, 289; Esri 2014b). The K(d) function is transformed and the L(d) function is calculated as $\sqrt{(K(d)/\pi)} - d$ (Bailey & Gatrell 1995, 94). For the L(d) function, the null hypothesis is zero (Sayer & Wienhold 2013, 78): when L(d) is greater than 0 there is clustering; in contrast, if L(d) is less than 0 there is regularity in the point distribution. The mathematical transformation of the Ripley's K-function performed in ArcGIS, however, is slightly different, and as a final result the expected index of a random pattern is equal to the input distance (Esri 2014b). In order to assess the pattern type, the observed L(d) value has to be compared with the value expected.

In each distance increment, the actual distribution is compared against a random distribution, and if the observed L(d) value (black curve in the graphs in Figs. 3.8 and 3.9) is greater than that expected (light gray curve), a clustered pattern is dominant at that scale of analysis. In turn, if this value is smaller than that expected, the distribution appears dispersed. Moreover, in order for the index of clustering or dispersion to be statistically significant, it must respectively be greater or smaller than the high or low level of the confidence interval (dark gray lines in the graphs). This index is indicated by the black curve (observed K): when this curve rises above the expected K curve (in light gray) and the high level of the confidence interval (superior dark gray line), the distribution is significantly clustered at that particular scale of analysis (distance). When, instead, the curve drops below the expected K curve and the low level of the confidence interval (lower dark gray line), the distribution is significantly dispersed at that scale of analysis.

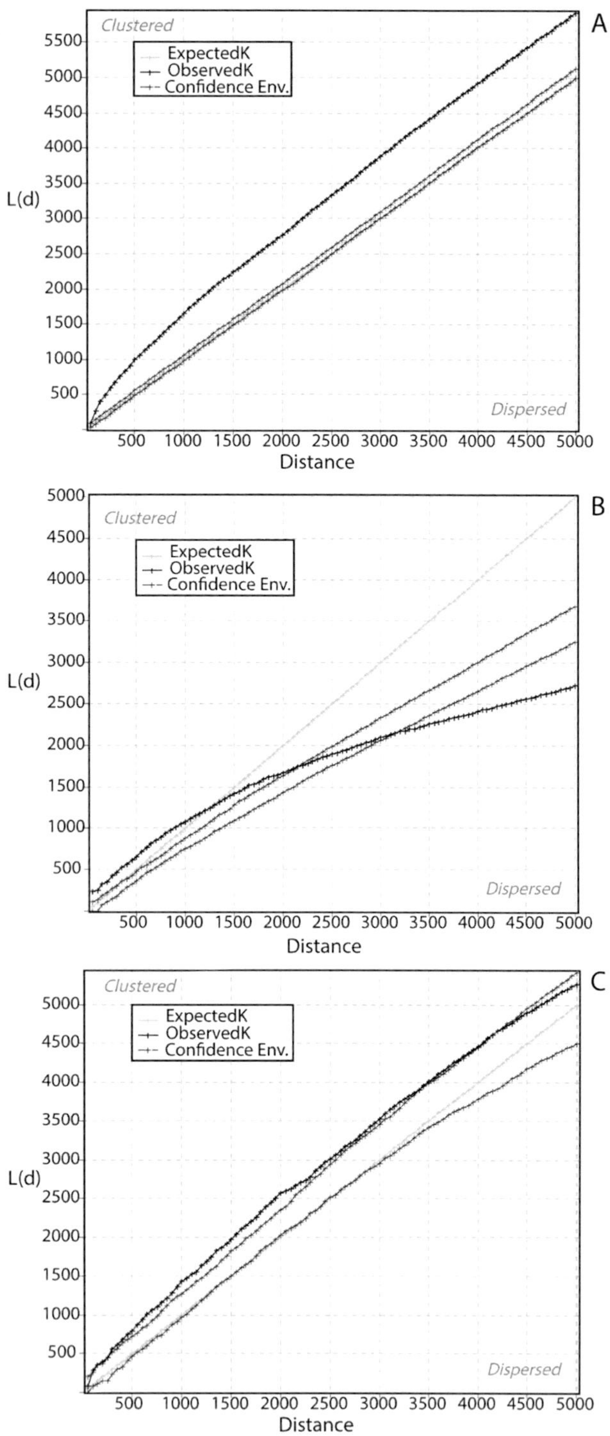

Fig. 3.8 Ripley's K-function analysis of the Hellenistic settlements. A) Ager Venusinus; B) Ager Cosanus; C) Ager Aeserninus.

In the Ager Venusinus and Aeserninus a statistically significant clustered pattern is predominant at most scales of analysis (graphs A and C in Fig. 3.8). In the Ager Venusinus, the clustering of Hellenistic sites is evident (the only distance increment in which clustering is not statistically significant is the first, from 0 to 50 m). In the Ager Aeserninus, a statistically significant nucleated pattern is attested from 100 m to 4.1 km and continues until 5 km (but for this latter distance it is not statistically significant). A different situation is encountered in the Ager Cosanus sample, where significant clustering appears only up to a maximum cumulative distance of 1.2 km. From 1.25 km to 5 km the sample seems more dispersed than a random distribution, and from 3.2 to 5 km this pattern even becomes statistically significant (graph B in Fig. 3.8).

Having established these clustered patterns with our pattern analysis, in a second, more detailed step, we tested for the regular dispersion of sites within these areas by zooming in on an ecologically uniform zone in each colonial territory. These three zones were selected on the basis of two criteria. First, these zones stand out for their extensive area of high site density. Second, the specific environmental and cultural characteristics of these zones are as homogeneous as possible. In this way, we limited the potential effect of attractive or repulsive socio-environmental factors on settlement distribution (*i.e.* first-order effects) while focusing on local distances between the settlements (*i.e.* second-order effects).

For the Ager Venusinus, the best candidate that meets these criteria is the wide plateau facing the colonial urban center, an area now called the Piani di Camera (Marchi & Sabbatini 1996, 111–115). This zone is close to the urban center, is characterized by uniformly smooth geomorphology and presents remarkably high settlement density and good survey-visibility conditions (Marchi & Sabbatini 1996, 107). In the Ager Cosanus, we selected a comparable area located close to the colonial town (Celuzza & Regoli 1982; Carandini *et al.* 2002, 137–138, 164–168). The valley floor of the Valle d'Oro is now used as arable land where the survey resulted in a relatively high site recovery rate (Carandini *et al.* 2002, 36–47). In the Ager Aeserninus, we focused on the western arm of the survey sample (Stek *et al.* 2015, 258–262), which is characterized by three river valleys and plateaus. The survey coverage in this zone is relatively high and more evenly distributed than in the other parts of the survey sample, average visibility is moderate, and the highest and most uniform site density is attested here.

The analysis was repeated on this selection of small eco-zones, looking at a maximum cumulative radius of 3000 m. As can be seen in graphs A and C in Fig. 3.9, no significant patterns are found on the Piani di Camera plateau and in the western transect. Sites seem to be located at random here. The analysis of Valle d'Oro, in contrast, showed significant nucleation up to 2.4 km (graph B in Fig. 3.9): generally, the pattern is clustered rather than randomly distributed. For the western transect of the Ager Aeserninus survey sample, there is some evidence of agglomeration found at a cumulative distance of 2.05 km, but this is not statistically significant, and we concluded that the site distribution here is random (graph C in Fig. 3.9).

To sum up, according to this targeted statistical analysis of the small eco-zones, the settlement pattern of the Piani di Camera in the Ager Venusinus and of the western transect in the Aeserninus survey sample may conform to the conventional model of Roman colonization assuming a sparse dispersion of rural farms (see also Marchi & Sabbatini 1996, 112 - 113). The opposite holds true for the Valle d'Oro in the Ager Cosanus, where the central cluster may indeed indicate a nucleated settlement strategy (but see Rathbone 1981; Celuzza & Regoli 1984; Carandini *et al.* 2002, 103–144 for a different interpretation).

3.5.1 MODELING A VILLAGE-BASED SETTLEMENT SYSTEM

In this section, we aim to further test the identified clustered pattern in several portions of the colonial territories. We compare the inter-village distances reconstructed from Bintliff's study of the evolution of Greek settlement systems with those of our potential early colonial villages. Bintliff (1999; 2000; 2009) developed a socio-ecological theory to investigate the origin of Greek city-states. He

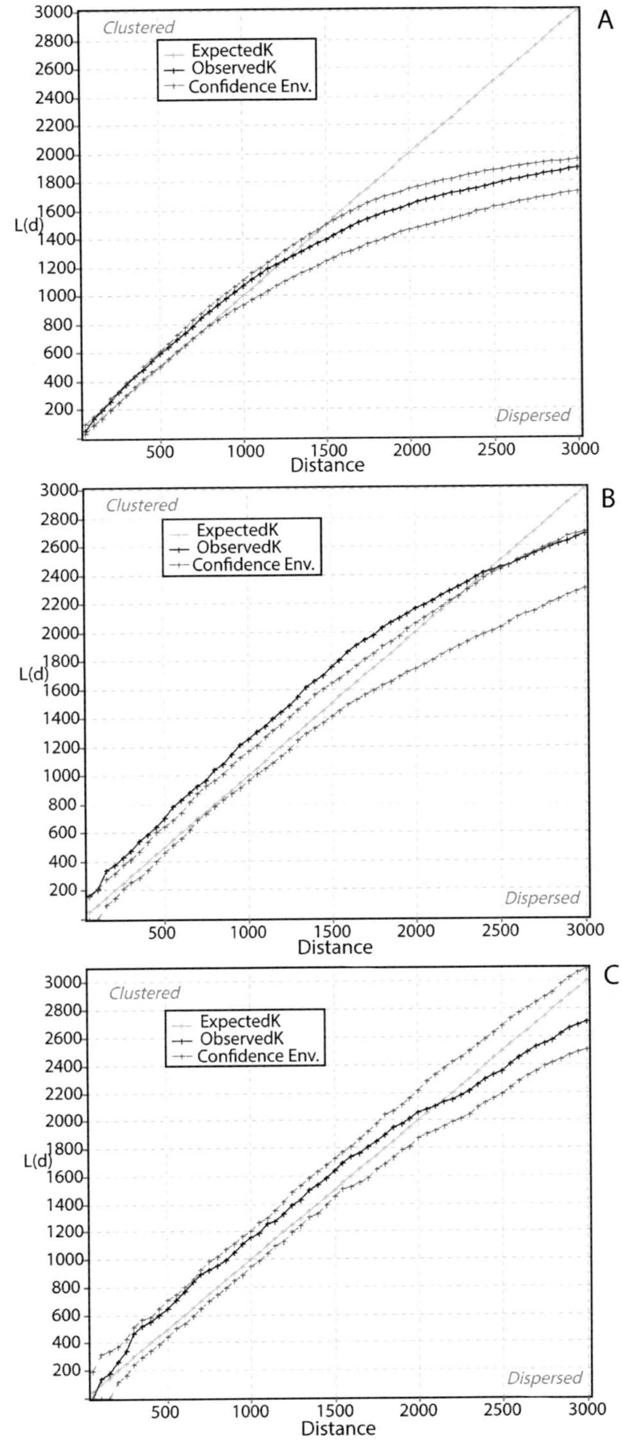

Fig. 3.9 Ripley's K-function analysis of the Hellenistic settlements. A) Piani di Camera (area: 7 sq km; number of sites: 78); B) Valle d'Oro (area: 15.65 sq km; number of sites: 47); C) western transect (area 29.62 sq km; number of sites: 39).

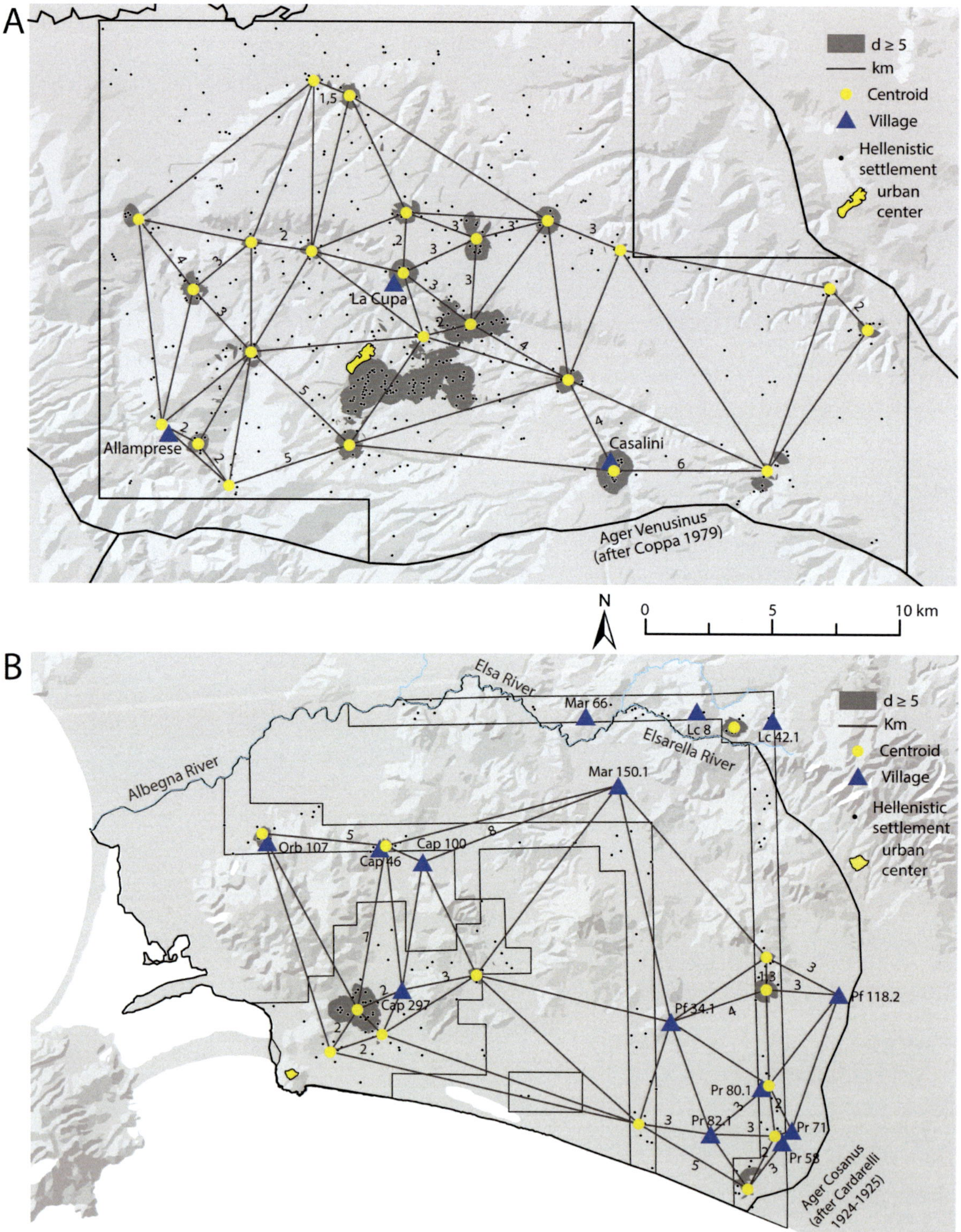

Fig. 3.10 Distances between centroids of scattered polygons (in gray) with site density ≥ five per sq km in the Ager Venusinus (A) and Cosanus (B). If a polygon is located less than 300 m from another one, they are merged and the centroid is placed centrally. Base map: hillshade elaboration of the 10 m-resolution DEM named TINITALY/01 (Tarquini *et al.* 2007; 2012).

argues that the formation of Greek city-states in the period *c.* 750–500 B.C. was triggered by the fission of scattered Dark Age settlements and the subsequent absorption of denser networks of Archaic settlements and territories by the most powerful village-states. In general, depending on the level of rural infill and demographic pressure, territorial catchments with radii of 5 km, 3–4 km, 2–3 km and 1–2 km can be recognized at different evolutionary stages in a village-based settlement model. On the basis of these estimations, it is possible to calculate Euclidean inter-village distances from 2 to 10 km. Bintliff (2000, 23) highlights that a sustainable rural infill in a Mediterranean landscape reveals village territories with radii of 2–3 km (4–6 km inter-village distance), and a 1–2 km catchment radius (2–4 km inter-village distance) may indicate noteworthy population pressure on the land. We tested these distance predictions against the distances between the potential early colonial villages we discerned.

First we selected potential early colonial nucleated settlements on which we can conduct this kind of comparative analysis. We selected the parts of the colonial landscape that displayed a very localized, high settlement density: cell clumps of at least 4–5 ha in area, with a density threshold equal to or higher than five sites per sq km (colored in gray in Fig. 3.10). We specify here that areas for which a random/dispersed pattern (conventional model) was ascertained with the Ripley's K function were not considered (*i.e.* the Piani di Camera plateau in the Ager Venusinus and the western transect of the Ager Aeserninus). This means that inter-village distances in the Ager Aeserninus could not be compared, since only one scattered high-site-density zone remains north of the city.

Both in the colonial territory of Venusia and Cosa, the distances between possible nucleated settlements match Bintliff's predictions. As displayed in Fig. 3.10A, the twenty-three zones of interest in the Ager Venusinus (their centroids are marked in yellow) have a minimum inter-distance of 1.5 km and a maximum inter-distance of 6 km from the nearest neighbor (average: 2.7 km). The most well-known pre-Roman villages (blue triangles, *i.e.* Casalini, Allamprese, La Cupa-Masseria La Gala) overlap these gray polygons of high site density. We are currently revisiting these (allegedly) Daunian and Samnite nucleated settlements in the context of the 'Landscapes of Early Roman Colonization' project (Stek & Pelgrom 2013; https://www.universiteitleiden.nl/en/research/research-projects/archaeology/landscapes-of-early-roman-colonization ; http://landscapesofearlyromancolonization.com/) to obtain additional information about a potential early colonial occupation. A detailed (re)examination of the black-gloss pottery collected at these villages very plausibly suggests continued occupation from the 4th to the 2nd century B.C. (Pelgrom *et al.* 2014). In the Ager Cosanus, as shown in Fig. 3.10B, the twelve zones of interest have a minimum inter-distance of 1.3 km and a maximum inter-distance of 5 km from the nearest neighbor (average: 2.8 km). The corner of the outmost transect probably belongs to the Ager Saturninus (Carandini *et al.* 2002, 159) and thus is not included in the analysis. Many of these zones coincide with or are very close to documented Etruscan and Roman villages (blue triangles, indicated by their acronyms [Carandini *et al.* 2002, 375–409]); if these are incorporated, they reduce the nearest-neighbor distance average to 2.5 km.

3.6 CONCLUSIONS

In this paper, we have tested the robustness of different settlement models that have been proposed for three mid-Republican colonies in Italy on the basis of regional field-survey projects. The existing datasets for Venusia, Cosa and Aesernia have been systematically subjected to point-density and pattern analysis. In all the three case studies, the results of the point-density analysis clearly indicate an enormous discrepancy from demographic reconstructions based on literary sources. Even if we accept the "best-case" scenario of early colonial occupation (by including all attested Hellenistic sites as possible early colonial sites) and select ideal visibility conditions for site discovery (modern arable areas and gentle slopes), three different survey teams, at three different moments and in three different landscapes (modern Basilicata, Tuscany and Molise), did not detect the supposedly numerous mid-Republican farms expected by traditional reconstructions. The pattern that emerges, instead, clearly indicates that the density of only a tiny portion of the territory is compatible with the expected number of sites. These small areas of high site-density are located primarily in the vicinity of the urban center (as at Venusia and Cosa), although this is not the rule (as in the case of Aesernia). In the rest of the survey samples, a scattering of localized high site densities suggests a patchier, clustered pattern.

It is clear that these results challenge conventional interpretations of colonial settlement organization in Republican Roman Italy. The significant differences in site density and pattern highlighted in our analysis are difficult to reconcile with conventional expectations of neatly partitioned territories, but fit the newly proposed polynuclear settlement scenario. It is true that some variants of the conventional model have suggested as well the presence of areas of nucleated settlement within colonial territories. In such cases, however, they have been interpreted as villages where the indigenous population settled (Coarelli 1991; Cornell 1995, 367; see Bradley 2006 on the inclusion of indigenous populations in colonies). In this view, the clustered pattern is reconciled with a scenario in which the native population is relocated in marginal zones of the *ager*, where it would have been allowed to continue settling in villages (Carandini *et al.* 2002, 108–110).

What then remains to be explained is where the colonists lived. If we consider the collected data to be largely representative of the early colonial settlement organization, there are two ways to answer this question. First, if we adhere to conventional interpretations, which consider densely and evenly distributed sites to be a diagnostic indicator of Roman colonial settlement, we must imagine an 'agro-town' (Garnsey 1979) in which a majority of colonists lived in the urban center and in a relatively small rural territory nearby, whereas the remaining indigenous population settled in clustered settlements farther away. Such a scenario, however, must assume that either Livy's demographic estimates are corrupt (*cf.* discussion in Pelgrom 2013) or include an additional population component, such as the colonists' family members or the indigenous population that continued living in their traditional villages (Torelli 1999, 94; Bradley 2006). Secondly, if, however, we accept Livy's colonial population numbers as a roughly correct representation of adult male colonial settlers, we must assume that the majority opted for a more nucleated settlement strategy (see also Torelli 1991, 22). They therefore may have colonized distant pockets of the conquered landscape at fairly standardized distances, which, as we have seen, match those set out by Bintliff (1999; 2000; 2009) in his village landscape model.

The latter explanation accords well with the recently proposed polynuclear colonial settlement strategy and further undermines the traditional model, which assumes that the indigenous people and the colonists would have followed radically different settlement patterns (villages versus evenly distributed single farms). Our analyses moreover show that the dispersed-versus-nucleated settlement dichotomy may be too limited and that other settlement rationales may be identified in the existing datasets. In the case of the Ager Venusinus, for instance, site density patterns correspond reasonably well to Von Thünen's density model of decreasing site concentration in progressive concentric bands around the urban market.

To sum up, the unilateral hinterland-city relationship projected by both the conventional colonial scenario (city provides political, administrative and defensive services to colonial farmers) and Von Thünen (farmers supply the city with their products) may not be the only way the countryside interacted. A complex network of powerful villages may indeed have had an important role in early colonial societal organization, and Roman rural settlement and economic strategies in colonial contexts may have been more diverse than previously assumed (see discussion in Stek 2018). The spatial-statistical analyses we have performed have provided an effective and methodologically sound way to test different settlement scenarios. At the same time, these analyses have enabled us to advance new, more flexible conceptual models for understanding colonial settlement strategies, which now may be tested further in the field and laboratory.

3.7 REFERENCES

Arnoldus-Huyzendveld, A. 2007. Le trasformazioni dell'ambiente naturale della pianura grossetana. Dati geologici e paleo-ambientali. In C. Citter & A. Arnoldus-Huyzendveld (eds.), *Archeologia urbana a Grosseto. Origine e sviluppo di una città medievale nella "Toscana delle città deboli". Le ricerche 1997–2005* I: *La città di Grosseto nel contesto geografico della bassa valle dell'Ombrone*, 62–71. Firenze: All'Insegna del Giglio.

Attwell, M. & Fletcher, M. 1985. A new technique for investigating spatial relationships: Significance testing. In A. Voorrips & S. H. Loving (eds.), *To pattern the past: Symposium on mathematical methods in archaeology, Amsterdam 1984*, 181–189. Strasbourg: Conseil de l'Europe.

Attwell, M. R. & Fletcher, M. 1987. An analytical technique for investigating spatial relationships. *Journal of Archaeological Science* 14 (1), 1–11. doi: 10.1016/S0305-4403(87)80002-X

Bailey, T. C. & Gatrell A. C. 1995. *Interactive spatial data analysis*. Harlow, Essex: Longman Scientific & Technical.

Barker, G. & Lloyd, J. (eds.) 1991. *Roman landscapes: Archaeological survey in the Mediterranean region*. London: British School at Rome.

Bevan, A. & Conolly, J. 2006. Multiscalar approaches to settlement pattern analysis. In G. Lock & B. Molyneaux (eds.), *Confronting scale in archaeology: Issues of theory and practice*, 217–234. New York: Springer.

Bevan, A. & Wilson, A. 2013. Models of settlement hierarchy based on partial evidence. *Journal of Archaeological Science* 40 (5), 2415–2427. doi: 10.1016/j.jas.2012.12.025

Bintliff, J. L. 1999. Chapter 13: Settlement and territory. In G. Barker (ed.), *Companion encyclopedia of archaeology*, 505–545. London: Routledge.

Bintliff, J. L. 2000. Settlement and territory: a socio-ecological approach to the evolution of settlement systems. In G. Bailey, R. Charles & N. Winder (eds.), *Human ecodynamics*, 21–30. Oxford: Oxbow Books.

Bintliff, J. L. 2009. Catchments, settlement chambers and demography: case studies and general theory in the Greek landscape from Prehistory to Early Modern times. In F. Favory & L. Nuninger (eds.), *ArchaeDyn. 7 Millenia of Territorial Dynamics. Settlement Pattern, Production and Trades from Neolithic to Middle Ages*, 107–117. Dijon: University of Burgundy.

Bintliff, J. L. & Sbonias, K. (eds.) 1999. *Reconstructing past population trends in Mediterranean Europe (3000 B.C.–A.D. 1800)*. Oxford: Oxbow Books.

Bintliff, J. L. & Snodgrass, A. M. 1985. The Cambridge/Bradford Boeotian expedition: The first four years. *Journal of Field Archaeology* 12, 123–161.

Bradley, G. 2006. Colonization and identity in Republican Italy. In G. J. Bradley & J. P. Wilson (eds.), *Greek and Roman colonization: Origins, ideologies and interactions*, 161–188. Swansea: Classical Press of Wales.

Brown, F. E. 1980. *Cosa: The making of a Roman town*. Ann Arbor: University of Michigan Press.

Cambi, F. 1999. Demography and Romanization in Central Italy. In J. L. Bintliff & K. Sbonias (eds.), *Reconstructing past population trends in Mediterranean Europe (3000 B.C.–A.D. 1800)*, 115–129. Oxford: Oxbow Books.

Carandini, A., Cambi, F., Celuzza, M. G. & Fentress, E. 2002. *Paesaggi d'Etruria. Valle dell'Albegna, Valle d'Oro, Valle del Chiarone, Valle del Tafone*. Roma: Edizioni di Storia e Letteratura.

Cardarelli, R. 1924–1925. Confini fra Magliano e Marsiliana; fra Manciano e Montauto, Scerpenna, Stachilagi; fra Tricosto e Ansedonia; fra Port'Ercole e Monte Argenatrio. *Maremma* 1–2.

Celuzza, M. G. 1984. La piccola proprietà. In S. Settis (ed.), *Misurare la terra: centuriazione e coloni nel mondo romano*, 158–160. Modena: Edizioni Panini.

Celuzza, M. G. & Regoli, E. 1982. La valle d'Oro nel territorio di Cosa. Ager Cosanus e Ager Veientanus a confronto. *Dialoghi di Archeologia* 4, 31–62.

Chisholm, M. 1968. *Rural settlement and land use. An essay in location*. London: Hutchinson.

Clark, P. J. & Evans, F. C. 1954. Distance to nearest neighbour as a measure of spatial relationships in populations. *Ecology* 35, 444–453. doi: 10.2307/1931034

Coarelli, F. 1991. I Sanniti a Fregellae. In *La romanisation du Samnium aux IIe et Ier siècles av. J.-C.*, 177–187. Centre Jean Bérard, Naples: Newprint s.p.a.

Conolly, J. & Lake, M. 2006. *Geographical Information Systems in archaeology*. Cambridge: Cambridge University Press.

Coppa, M. 1979. Formazione della struttura territoriale dell'agro di Venosa. In M. Coppa & G. Alvisi (eds.), *Fotografia aerea e storia urbanistica*, 119–128. Roma: Edigraf.

Cornell, T. 1995. *The Beginnings of Rome: Italy and Rome from the Bronze Age to the Punic Wars (c. 1000–264 B.C.)*. London: Routledge.

De Neeve, P. W. 1984. *Peasants in peril: Location and economy in Italy in the second century B.C.* Amsterdam: JC Gieben.

Dixon, P. M. 2002. Ripley's K function. In A. H. El-Shaarawi & W. W. Piegorsch (eds.), *Encyclopedia of environmetrics*, 1796–1803. New York: John Wiley & Sons.

Dodson, R. F. 1991. *VT/GIS: the Von Thünen GIS package*. Santa Barbara: National Center for Geographic Information and Analysis, University of California at Santa Barbara.

Eastman, J. R. 2012. *IDRISI Selva manual (Manual Version 17.01)*. Worcester, MA: Clark Labs for Cartographic Technology and Geographic Analysis, Clark University.

Esri, 2014a. Point Density (Spatial Analyst). In *ArcGIS 10.2.2 Help*.

Esri, 2014b. Multi-distance spatial cluster analysis (Ripley's K Function) (Spatial Statistics). In *ArcGIS 10.2.2 Help*.

FAO, 2006. *Guidelines for soil description*. Rome: Food and Agriculture Organization of the United Nations.

Feiken, H. 2014. *Dealing with biases: Three geo-archaeological approaches to the hidden landscapes of Italy*. Eelde, Netherlands: Barkhuis.

Fentress, E. 2000. What are we counting for? In R. Francovich, H. Patterson & G. Barker (eds.), *Extracting meaning from ploughsoil assemblages*, 45–52. Oxford: Oxbow Books.

Fentress, E. 2009. Peopling the countryside: Roman demography in the Albegna valley and Jerba. In A. Bowman & A. Wilson (eds.), *Quantifying the Roman economy. Methods and problems*, 127–163. Oxford: Oxford University Press.

Garnsey, P. 1979. Where did Italian peasants live? *PCPhS* 25, 1–26.

Given, M. 2004. Mapping and manuring: Can we compare sherd density figures? In S. E. Alcock & J. F. Cherry (eds.), *Side-by-side survey: Comparative regional studies in the Mediterranean World*, 13–21 Oxford: Oxbow Books.

Goodchild, H. 2007. *Modelling Roman agricultural production in the Middle Tiber valley, Central Italy*. PhD dissertation, University of Birmingham: eTheses Repository http://etheses.bham.ac.uk/175/

Goodchild, H. & Witcher, R. 2010. Modelling the agricultural landscapes of Republican Italy. In J. Carlsen & E. Lo Cascio (eds.), *Agricoltura e scambi nell'Italia tardo Repubblicana*, 187–220. Roma & Bari: Edipuglia.

Grotewold, A. 1959. Von Thünen in retrospect. *Economic Geography* 35 (4), 346–355. doi: 10.2307/142467

Haggett, P., Cliff, A. D. & Frey, A. P. 1977. *Locational analysis in human geography*. London: Arnold.

Hodder, I. & Orton, C. 1976. *Spatial analysis in archaeology*. Cambridge: Cambridge University Press.

Horden, P., & Purcell, N. 2000. *The corrupting sea: A study of Mediterranean history*. Oxford: Wiley-Blackwell.

Hudson, J. C. 1969. A location theory for rural settlement. *Annals of the Association of American Geographers* 59 (2), 365–381.

doi: 10.1111/j.1467-8306.1969.tb00676.x

Jameson, M. H., Runnels C. N. & Van Andel T. H. 1994. *A Greek countryside. The Southern Argolid from Prehistory to the present day*. Stanford: Stanford University Press.

Judge, W. J. & Sebastian, L. (eds.) 1988. *Quantifying the present and predicting the past: Theory, method and application of archaeological predictive modeling*. Denver: Bureau of Land Management.

Kamermans, H. 2000. Land evaluation as predictive modelling: a deductive approach. In G. Lock (ed.), *Beyond the map: Archaeology and spatial technologies*, 124–146. Amsterdam: IOS Press.

Kintigh, K. W. & Ammerman, A. J. 1982. Heuristic approaches to spatial analysis in archaeology. *American Antiquity* 47, 31–63. doi: 10.2307/280052

Kvamme, K. L. 1990. The fundamental principles and practice of predictive archaeological modelling. In A. Voorrips (ed.), *Mathematics and information science in archaeology: A flexible framework*, 257–295. Bonn: Holos.

Launaro, A. 2011. *Peasants and slaves. The rural population of Roman Italy (200 B.C. to A.D. 100)*. Cambridge: Cambridge University Press.

Lock, G. & Molyneaux, B. L. (eds.) 2006. *Confronting scale in archaeology: Issues of theory and practice*. New York: Springer.

Marchi, M. L. 2010. *Ager Venusinus II: (IGM 175 II SO; 187 I NO; 187 I SE; 188 IV NO; 188 IV SO)* (Forma Italiae 43). Firenze: L.S. Olschki.

Marchi, M. L. & Sabbatini, G. 1996. *Venusia (IGM 187 I NO / I NE)* (Forma Italiae 37). Firenze: L.S. Olschki.

Millett, M. 1991. Pottery: Population or supply pattern? The Ager Tarraconensis approach. In G. Barker & J. Lloyd (eds.), *Roman landscapes: Archaeological survey in the Mediterranean region*, 18–26. London: British School at Rome.

Morley, N. 1996. *Metropolis and hinterland. The city of Rome and the Italian economy 200 B.C.–A.D. 200*. Cambridge: Cambridge University Press.

Orton, C. 2004. Point pattern analysis revisited. *Archeologia e Calcolatori* 15, 299–315.

Palmisano, A. 2013. Zooming patterns among the scales: A statistics technique to detect spatial patterns among settlements. In G. Earl, T. Sly, P. Chrysanthi., P. Murrieta-Flores, C. Papadopoulos, I. Romanowska & D. Wheatley (eds.), *CAA 2012: Archaeology in the Digital Era*, 348–356. Amsterdam: Amsterdam University Press.

Patterson, J. R. 2004. City, territory and metropolis: The case of the Tiber valley. In H. Patterson (ed.), *Bridging the Tiber: Approaches to regional archaeology in the middle Tiber valley*, 61–73. London: British School at Rome.

Pelgrom, J. 2008. Settlement organization and land distribution in Latin colonies before the Second Punic War. In L. De Ligt & S. J. Northwood (eds.), *People, land and politics. Demographic developments and the transformation of Roman Italy 300 B.C.-A.D. 14*, 317–356. Leiden: Brill.

Pelgrom, J. 2012. *Colonial landscapes. Demography, settlement organization and impact of colonies founded by Rome (4th–2nd centuries B.C.)*. PhD Dissertation, Leiden University: Leiden University Repository.

Pelgrom, J. 2013. Population density in mid-Republican Latin colonies: A comparison between text-based population estimates and the results from survey archaeology. *Atlante Tematico di Topografia Antica* 23, 73–84.

Pelgrom, J. 2014. Roman colonization and the city-state model. In T. D. Stek & J. Pelgrom (eds.), *Roman Republican colonization. New perspectives from archaeology and ancient history*, 70–84. Rome: Palombi.

Pelgrom, J., Marchi, M. L., Cantoro, G., Casarotto, A., Hamel, A., Lecce, L.,García-Sánchez, J. & Stek, T. D. 2014. New approaches to the study of village sites in the territory of Venosa in the Classical and Hellenistic periods. *Agri Centuriati* 11, 31–60.

Perles, C. 2001. *The early Neolithic in Greece: the first farming communities in Europe*. Cambridge: Cambridge University Press.

Rathbone, D. W. 1981. The development of agriculture in the 'Ager Cosanus' during the Roman Republic: Problems of evidence and interpretation. *Journal of Roman Studies* 71, 10–23. doi: 10.2307/299493

Rathbone, D. W. 2008. Poor peasants and silent sherds. In L. De Ligt & S. J. Northwood (eds.), *People, land, and politics. Demographic developments and the transformation of Roman Italy, 300 B.C.–A.D. 14*, 305–332. Leiden: Brill.

Ripley, B. D. 1976. The second-order analysis of stationary point processes. *Journal of Applied Probability* 13 (2), 255–266. doi: 10.1017/S0021900200094328

Roberts, B. K. 1996. *Landscapes of settlement: prehistory to the present*. London: Routledge.

Sabbatini, G. 2001. *Ager Venusinus I: Mezzana del Cantore, IGM 175 II SE* (Forma Italiae 40). Firenze: L. S. Olschki.

Salmon, E. T. 1969. *Roman colonization under the Republic*. London: Thames and Hudson.

Sambucini, V., Marinosci, I., Bonora, N. & Chirici, G. 2010. *La realizzazione in Italia del progetto Europeo Corine Land Cover 2006*. Rome: ISPRA.

Sayer, D., & Wienhold, M. 2013. A GIS-investigation of four Early Anglo-Saxon cemeteries: Ripley's K-function analysis of spatial groupings amongst graves. *Social Science Computer Review* 31 (1), 71–89. doi: 10.1177/0894439312453276

Shennan, S., Gardiner, J. & Oake, M. 1985. *Experiments in the collection and analysis of archaeological survey data: the East Hampshire survey*. Sheffield: University of Sheffield.

Stek, T. D. 2008. *Sanctuary and society in Central-Southern Italy. a study into cult places and cultural change after the Roman conquest of Italy*. PhD dissertation, University of Amsterdam: UvA-DARE (Digital Academic Repository).

Stek, T. D. 2009. *Cult places and cultural change in Republican Italy: a contextual approach to religious aspects of rural society after the Roman conquest*. Amsterdam: Amsterdam University Press.

Stek, T. D. 2014. The city-state model and Roman Republican colonization: sacred landscapes as a proxy for colonial socio-political organization. In T. D. Stek & J. Pelgrom (eds.), *Roman Republican colonization. New perspectives from archaeology and ancient history*, 87–105. Rome: Palombi.

Stek, T. D. 2018. Early Roman colonization beyond the Romanizing agro-town: village patterns of settlement and highland exploitation in Central Italy. In B. S. Düring & T. D. Stek (eds.), *The Archaeology of Imperial Landscapes: A Comparative Study of Empires in the Ancient Near East and Mediterranean World*, 145-172. Cambridge: Cambridge University Press.

Stek, T. D., Modrall, E. B., Kalkers, R. A. A., van Otterloo, R. H. & Sevink, J. 2015. An early Roman colonial landscape in the Apennine mountains: landscape archaeological research in the territory of Aesernia (Central-Southern Italy). *Analysis Archaeologica. Studies in Western Mediterranean Archeology* 1, 229–291.

Stek, T. D. & Pelgrom, J. 2013. Landscapes of Early Roman Colonization: Non-urban settlement organization and Roman expansion in the Roman Republic (4th–1st centuries B.C.). *Tijdschrift voor Mediterrane Archeologie* 50, 87.

Tarquini, S., Isola, I., Favalli, M., Mazzarini, F., Bisson, M., Pareschi, M. T. & Boschi, E. 2007. TINITALY/01: A new Triangular Irregular Network of Italy. *Annals of Geophysics* 50 (3), 407–425. doi: 10.4401/ag-4424

Tarquini, S., Vinci, S., Favalli, M., Doumaz, F., Fornaciai, A. & Nannipieri, L. 2012. Release of a 10-m-resolution DEM for the Italian territory: Comparison with global-coverage DEMs and anaglyph-mode exploration via the web. *Computers & Geosciences* 38 (1), 168–170. doi:10.1016/j.cageo.2011.04.018.

Terrenato, N. 2000. The visibility of sites and the interpretation of field survey results: towards an analysis of incomplete distributions. In R. Francovich, H. Patterson & G. Barker (eds.), *Extracting meaning from ploughsoil assemblages*, 60–71. Oxford: Oxbow Books.

Terrenato, N. & Ammerman, A. J. 1996. Visibility and site recovery in the Cecina valley survey, Italy. *Journal of Field Archaeology* 23 (1), 91–109.

Thornton, P. K. & Jones, P. G. 1998. A conceptual approach to dynamic agricultural land-use modelling. *Agricultural Systems* 57 (4), 505–521. doi: 10.1016/S0308-521X(98)00005-5

Torelli, M. 1991. La fondazione di Venosa nel quadro della romanizzazione dell'Italia meridionale. In M. Salvatore (ed.), *Il Museo archeologico nazionale di Venosa*, 17–26. Matera: IEM editrice.

Torelli, M. 1999. *Tota Italia: Essays in the cultural formation of Roman Italy*. Oxford: Clarendon Press.

Toynbee, A. J. 1965. *Hannibal's legacy. 1, Rome and her neighbours before Hannibal's entry*. London: Oxford University Press.

Van Leusen, P. M. 2002. *Pattern to process: Methodological investigations into the formation and interpretation of spatial patterns in archaeological landscapes*. PhD Dissertation, University of Groningen: University of Groningen Research Database.

Van Leusen, M. & Kamermans, H. (eds.) 2005. *Predictive modelling for archaeological heritage management: A research agenda*. Amersfoort: ROB.

Van Leusen, M., Pizziolo, G. & Sarti, L. (eds.) 2011. *Hidden landscapes of Mediterranean Europe. Cultural and methodological biases in Pre- and Protohistoric landscape studies*. BAR International Series 2320. Oxford: Archaeopress.

Verhagen, P. 2007. *Case studies in archaeological predictive modelling* (ASLU 14). Leiden: Leiden University Press.

Vita-Finzi, C. & Higgs, E. S. 1970. Prehistoric economy in the Mount Carmel area of Palestine: Site catchment analysis. *Proceedings of the Prehistoric Society* 36, 1–37.

Von Thünen, J. H. 1966. *Von Thünen's 'Isolated state': An English edition of Der Isolierte Staat*. C.M. Wartenberg, trans. P. Hall (ed.) Oxford: Pergamon.

Wheatley, D. & Gillings, M. 2002. *Spatial technology and archaeology. The archaeological applications of GIS*. London & New York: Taylor & Francis.

Winter-Livneh, R., Svoray, T. & Gilead, I. 2010. Settlement patterns, social complexity and agricultural strategies during the Chalcolithic period in the Northern Negev, Israel. *Journal of Archaeological Science* 37 (2), 284–294. doi: 10.1016/j.jas.2009.09.039

Witcher, R. E. 2008. (Re)surveying Mediterranean rural landscapes: GIS and legacy survey data, *Internet Archaeology* 24. doi: 10.11141/ia.24.2.

Witcher, R. E. 2009. The middle Tiber valley in the Imperial period. In F. Coarelli & H. Patterson (eds.), *Mercator Placidissimus. The Tiber valley in antiquity. New research in the Upper and Middle river valley*, 467–486. Rome: Quasar.

Witcher, R. E. 2011. Missing persons? Models of Mediterranean regional survey and ancient populations. In A. Bowman & A. Wilson (eds.), *Settlement, urbanization and population*, 36–75. Oxford: Oxford University Press.

Zipf, G. K. 1949. *Human behavior and the principle of least effort: An introduction to human ecology*. Cambridge, MA: Addison-Wesley Press.

Chapter 4. Assessing visibility and geomorphological biases in regional field surveys: the case of Roman Aesernia[15]

Abstract
Archaeological field survey data can be biased by many factors, such as ground visibility conditions (*e.g.* vegetation, plowing) and geomorphological processes (erosion, deposition). Both visibility and geomorphological factors need, therefore, to be assessed when patterns of settlement and location preferences are inferred from survey data. Although both factors have been taken into account in a variety of fieldwork projects and studies, their combined effects remain hard to predict. In this paper, we aim to address this issue by presenting a visualization method that helps evaluating in combination the possible visibility and geomorphological effects in regional, site-oriented field surveys. Capitalizing on first-hand data on both archaeology and soil types produced by the recent Leiden University field survey project in the area of Isernia (Roman Aesernia, Central-Southern Italy), we propose a combined application of statistical tests and geo-pedological analysis to assess the extent and scale of the main biases possibly affecting the interpretation of the ancient settlement organization. Translating both sets of biases into GIS maps, we indicate the *likelihood* that negative field survey observations (absence of sites), in specific parts of the landscape, are genuine or rather distorted by biasing factors. The resulting 'archaeological detectability' maps allow researchers to formally highlight critical surveyed zones where the recording of evidence is likely unreliable, and thus provide a filter through which archaeologists can calibrate their interpretations of field survey datasets.

4.1 INTRODUCTION

Survey data are essentially fragmentary and biased by visibility factors and geomorphological processes. The impact of surface visibility on the recognizability of archaeological material at the surface has been a central debate in field survey archaeology in the last decades (*e.g.* De Guio 1985; Allen 1991; Terrenato & Ammerman 1996; Francovich *et al.* 2000; van Leusen 2002; Given 2004; Terrenato 2004; van Leusen *et al.* 2011). At present, general agreement exists that our view of the past as offered by field survey data is critically distorted by many factors (Banning 2002, 39-79 for a summary of these factors) and that we should be very cautious using these data uncritically (Fentress 2000). As a way to get around some of these biases, various correction methods have been proposed (*e.g.* Nance 1983; Shennan *et al.* 1985; Verhoeven 1991; van Leusen 1996; 2001; Gilling & Sbonias 1999; Terrenato 2000). Such studies, moreover, show that not only may surface visibility conditions impact heavily on site recovery rates, but that it is also fundamental to take into account the role of Holocene erosional and depositional processes, before attempts are made to infer settlement patterns

15 This Chapter corresponds to the article "Assessing visibility and geomorphological biases in regional field surveys: The case of Roman Aesernia" by Anita Casarotto, Tesse D. Stek, Jeremia Pelgrom, Ruud H. van Otterloo & Jan Sevink, originally published (17 July 2017) in *Geoarchaeology: An International Journal,* volume 33 issue 2 (publisher Wiley) as an Open Access article distributed under the terms of the Creative Commons Attribution-NonCommercial-NoDerivatives License (https://creativecommons.org/licenses/by-nc-nd/4.0/). This article is accessible through this link: https://doi.org/10.1002/gea.21627
The authors of the article wish to thank the members of the LERC project (Landscapes of Early Roman Colonization), and especially the survey team leaders Rogier Kalkers and Dr. Jesús García Sánchez and all the other team leaders, field walkers, and pottery specialists who took part in the various LERC campaigns in Isernia, for their hard work and enthusiasm that permitted the collection of important data for the study of ancient settlement dynamics in this territory. We are also grateful for the support to Soprintendenza Archeologia Belle Arti e Paesaggio del Molise, and to various local municipalities and institutions in Molise. The first author also wishes to thank Prof. Carlo Citter for the stimulating discussion on settlement patterns and settlement "emptiness" in landscape archaeology.

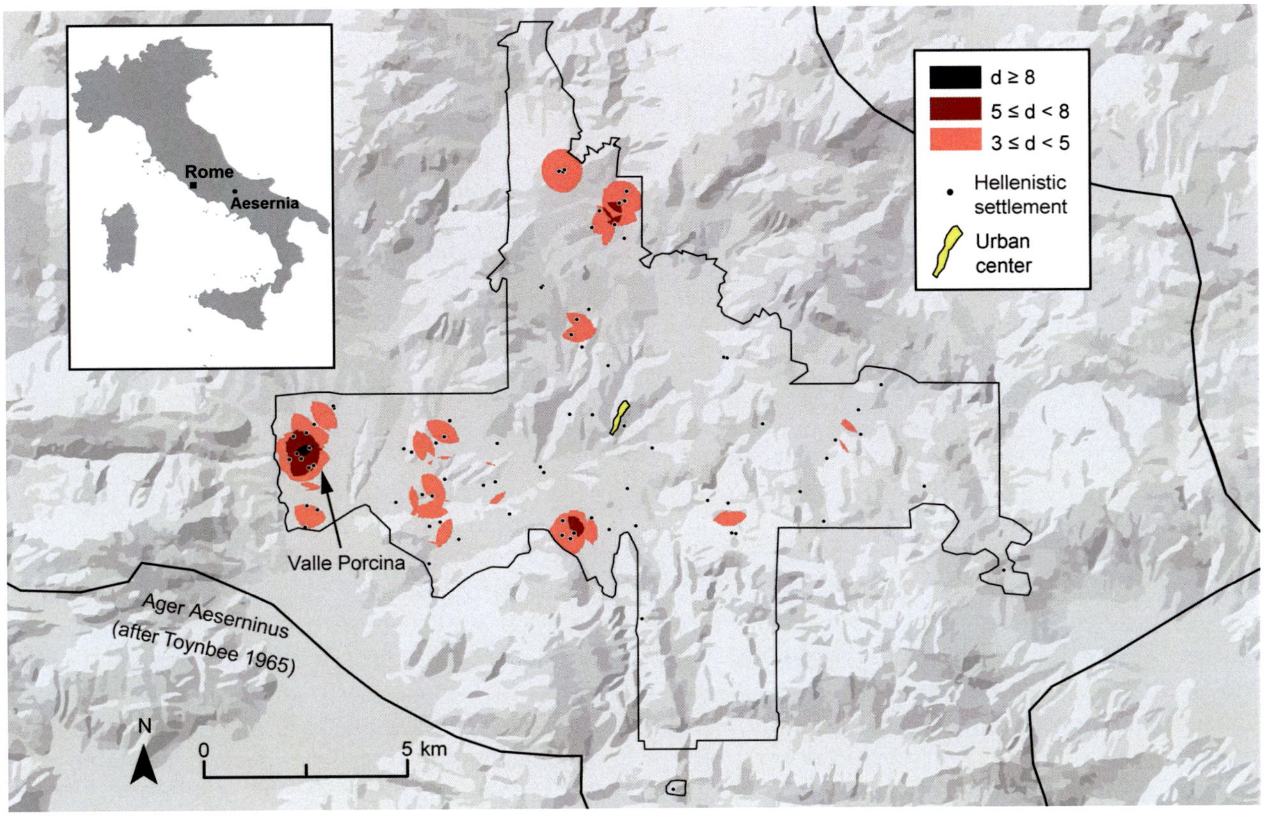

Fig. 4.1 Distribution of the Hellenistic settlements registered by the LERC survey project in the survey sample area. The palette highlights areas of different site density (d): d indicates the number of sites found in a circle of 1 sq km from each location (20 by 20 m) of the study area. Base map: graphic elaboration of the shaded relief calculated from the 10 m-resolution DEM named TINITALY/01 (Tarquini *et al.* 2007; 2012; Tarquini & Nannipieri 2017).

from field survey data (for a discussion on this theme for Mediterranean landscapes see Vita-Finzi 1969; Potter 1976; Sevink 1985; Bintliff 1992; Leonardi 1992a; Barker 1995a; Brown 1997, 237-248; 1999; Bintliff 2000; Vermeulen & De Dapper 2000; Feiken 2014; Walsh 2014; Koopman *et al.* 2016).

Through a systematic analysis of the most frequent methodological biases affecting the discovery of archaeological material during field survey, and using results from archaeological and soil research in the area of Isernia (Molise, Italy), we aim to assess the extent and scale of these biases in this region. Our test case regards the territory of the ancient settlement of Aesernia, where during the Roman conquest of Italy a colony was established by Rome in 263 B.C.. Since 2011 a large-scale archaeological project has been carried out in the territory of this ancient town, mapping and (re-)considering both archaeological and soil characteristics of the area (Stek *et al.* 2015).

In the current debate on Roman colonization, there are divergent views about site densities and settlement organization in early Roman colonial territories (see Casarotto *et al.* 2016 with further references, *i.e.* Chapter 3 of this book). The assessment of potential biases in the archaeological surface record, therefore, is becoming particularly pressing because it may eventually disclose which among these theories is the most plausible one. In the specific case of the territory of Aesernia, Stek *et al.* (2015) noted a variegated settlement arrangement in the distribution of Hellenistic and early colonial sites, characterized by long tracts of empty space in-between localized concentrations of settlements, although a regularly dispersed settlement pattern was discerned in a portion of the western part of the survey sample area (see Fig. 4.1). If this variegated pattern of colonial sites reflected historical reality, it would differ considerably from conventional models of Roman colonial territorial organization. Considering the high impact that such an unexpected pattern of colonial settlement has for historical debates on Roman colonization, it becomes crucial to thoroughly assess the possibility that the recorded configuration is actually patterned by biasing factors. By combining archaeological survey data, field observations, and soil information, we develop a method to test whether the recorded early colonial site distribution in the territory of Aesernia is the result of visibility and geomorphological biases, and if so, to what extent.

4.2 DATA

Our analysis capitalizes on the recently collected dataset of archaeological sites registered by the LERC team (Landscapes of Early Roman Colonization project, http://www.universiteitleiden.nl/en/research/research-projects/archaeology/landscapes-of-early-roman-colonization; https://landscapesofearlyromancolonization.com/) from 2011 to 2015 in the territory of the ancient colony of Aesernia (Stek *et al.* 2015). The project was first started with an EU Marie-Curie fellowship (FP7) granted to the second author and was based at Glasgow University, but in 2012 it moved to Leiden University and since then has been funded by the Netherlands Organization for Scientific Research (NWO) and the Royal Netherlands Institute in Rome (KNIR). The survey can be qualified as a regional, site orientated survey: within a surveyed area of 1886.031 ha, all encountered scatters of archaeological material at the surface were mapped with a GPS, using a threshold for site detection of five sherds per sq m. Survey teams, consisting generally of five walkers spaced ten meters apart, investigated each accessible field unit (in total 6116 units, see Stek *et al.* 2015, 255 – 257 for further details on the size of these units). In total 99 archaeological sites were identified, of which 81 are interpreted as probable/possible Hellenistic settlements (Fig. 4.1).

The other primary dataset used in this analysis consists of information provided by two soil surveys of the area around Isernia (Koopmans 1980; van Otterloo 1981; van Otterloo & Sevink 2016). This is used to assess the occurrence of possible recent gradational processes of erosion and deposition at a regional scale. The original soil surveys were carried out by the Laboratory for Physical Geography and Soil Science of the University of Amsterdam (Koopmans 1980; van Otterloo 1981) and a summary map was published later on (van Otterloo & Sevink 1983). The soil map by Koopmans was originally at scale 1:50,000 and covers most of the territory surveyed by the LERC team around Isernia, whereas Van Otterloo's soil map is at scale 1:25,000 and part of it covers the extreme portion of the west survey transect (published in Stek *et al.* 2015, 290-291). These maps were recently reviewed and checked in the field by van Otterloo and Sevink as part of the LERC project in order to produce a single, integrated map (van Otterloo & Sevink 2016). This integrated map is at scale 1:25,000 and was obtained through a cross-check of the previously collected soil information and maps (*i.e.* Koopmans 1980; van Otterloo 1981) with newly available higher-detail topographic maps (CTRs maps – Carta Tecnica Regionale, 1:5000). The part of this updated soil map covering the area under investigation in this paper is shown in Fig. 4.3.

4.3 METHODS

Survey visibility factors and geomorphological processes operate at different spatial and temporal levels, and can interrelate in intricate ways. Modeling their combined effects with respect to site detection in order to predict their potential impact on the survey record is therefore complex. For example, soil erosion peaks drastically on arable land, lacking a protective arboreal mantle (water erosion is induced) and seasonally disturbed by tillage activities (Torri *et al.* 2006), but a cleared or plowed field usually offers better visibility for the survey (and, thus, a higher discovery expectancy) than a vegetated area. Because of these complex interrelationships, in this paper we first discuss survey visibility factors and geomorphological processes separately and only in a final phase consider their combined effects on site distributions. To maintain a defined chronological focus we only consider the early Roman or Hellenistic sites as attested by black gloss pottery (*c.* 350 – 50 B.C.).

First, we assessed the role played by visibility conditions in favoring or preventing site discovery in the walked field units of the Aesernia survey. We focused on the physical characteristics of the modern landscape (*e.g.* vegetation cover, tillage status, or land use) and their correlations (if any) with site discovery. We considered whether the recording of sites by field walkers may have depended on physical constraints by means of statistical tests (*i.e.* Chi-square, Kolmogorov-Smirnov and Atwell-Fletcher tests). These tests highlight visibility categories where the number of recorded sites is significantly higher or lower, thus offering an indication of possibly biased samples.

As part of this analysis, a Multi Criteria Evaluation (MCE) was implemented in IDRISI GIS (Selva edition) (Eastman 2012) in order to produce a map indicating favorable units for survey visibility. The MCE is a decision support tool frequently applied in archaeological predictive modeling for location preference analysis of past settlements (*e.g.* Goodchild 2007; Di Zio & Barnabei 2009; Casarotto *et al.* 2011). It combines different criteria (predictors) and, depending on the problem at issue, shows the most suitable choice to be taken amongst many alternative solutions. Since most predictive models aim at identifying the most attractive landscape features for settlement in antiquity, this choice usually regards the most suitable locations for ancient settlements or past agricultural land use (see the discussion in De Guio 2015). In our case, instead, we wanted to choose the most suitable units (choice) for survey visibility (problem) according to a set of visibility factors (criteria/predictors).

The most influential visibility factors for archaeological detection (which we previously established through quantitative tests) were used in this model as explanatory variables. This means that, according to their weight of importance, they proportionally contributed to the construction of a suitability map showing which surveyed units are expected to offer a high probability of encountering a site based on the favorable visibility conditions. Such a predictive map helps us to highlight where absence of evidence may indeed reflect a real evidence of absence of settlement in antiquity (but see next section on geomorphological biases), and conversely which units may have instead yielded unreliable information, and thus where absence of evidence can likely be explained by adverse visibility conditions.

The second approach aimed at investigating the extent to which Late Holocene erosional and depositional processes may have destroyed or obscured Hellenistic sites. If degraded or buried, such sites could not have been detected in the topsoil during field surveys, even though survey visibility conditions were optimal. The absence of recorded sites there may not be reliable and geomorphological filtering is necessary before attempts can be made to interpret the data in historical terms. This is a well-known issue that has attracted much attention of scholars working in the Mediterranean world, above all ever since New Archaeology stimulated the analysis of depositional and post-depositional formative processes of the archaeological record (*e.g.* Clarke 1968; Schiffer 1987; Leonardi 1992b).

Geo-pedological investigations are increasingly being carried out in Mediterranean archaeological studies, not only for land evaluation analyses of ancient agricultural practices (*e.g.* Finke *et al.* 1994; Barker 1995b; Kamermans 2000; Van Joolen 2003; Goodchild 2007; Kamermans & Sevink 2009; Citter & Arnoldus-Huyzenveld 2011; Brown & Walsh 2017) but also for assessing geomorphological biases in the survey results. As regards the latter, sophisticated computer-based simulation models have been recently developed in order to study and quantify long-term sedimentation and erosion rates possibly affecting the archaeology (*e.g.* Zwertvaegher 2012; Feiken 2014, 125-162) or long-term soil degradation related to ancient agropastoral land-uses (*e.g.* Barton *et al.* 2010). However, the most widely applied methods for assessing archaeological preservation potential are those based on more general landscape classification procedures of geological and soil maps (*e.g.* Leonardi 1992a; Ebert & Singer 2004; Arnoldus-Huyzendveld 2007, 2011; Feiken 2014, 57-122; see the discussion in Sevink 1985). These methods are much more intuitive than the previous ones and more appropriate for a regional analysis that aims at testing sedimentation or erosion affecting large-scale settlement patterns. Here, we apply as well a particular landscape classification procedure of soil maps to assess the Late Holocene gradational effects on our Hellenistic settlement distribution.

More specifically, we analyzed the information provided by the soil map of the area (Koopmans 1980; van Otterloo 1981; van Otterloo & Sevink 2016) to gain understanding of the most recent sedimentation or erosive processes that may have occurred after the Hellenistic period. On the basis of this evaluation, we established the *detectability* (that is, the probability of recording sites, also known as preservation potential) of Hellenistic-period archaeology in each soil unit. We used the Attwell-Fletcher test of association (1985; 1987) for analyzing whether geomorphological biases may have affected the detection of Hellenistic sites, and to what extent.

As a last step, we combined the results of the visibility and geomorphological analyses in order to obtain a comprehensive detectability map of the walked survey units: to each unit, a pair of values was appointed indicating the probabilities that the site sample (or the vacuum) recorded in that unit is representative and, thus, reliable. This was done in IDRISI through a cross-tabulation that displays, for each surveyed unit, the different Hellenistic visibility and preservation potentials.

4.4 EVALUATING SURFACE VISIBILITY BIASES

In this first section, we focus on the physical obstacles at the surface that can affect the results of field surveys, such as vegetation cover or modern land use. During the field surveys carried out in Isernia, the visibility conditions of each walked field unit were registered systematically. Precise indications about tillage status, soil humidity, shadow conditions, stoniness, presence of recent material at the surface, vegetation cover, and land use were recorded on a scale from one (low) to five (high) in a database and reproduced in GIS as georeferenced vector files. Such indications are particularly precious for the purpose of the present analysis, since they allowed us to highlight possible significant correlations between site recovery rates and survey visibility conditions.

Using predictive modeling techniques, we modeled the aptitude for discovery of archaeological sites by modern field walkers in relation to more or less favorable visibility conditions. This analysis helped us to point out those surveyed units where, despite the optimal visibility, site discovery did not happen. The emptiness attested there may be due to either past constraints against settlement (thus ancient people chose to avoid those locations) or to a possible geomorphological bias affecting the preservation of sites, which we will address separately in section 4.5.

Before proceeding, however, we needed to take into account that different types of sites are variably visible in the survey record according to the periods and regions (Sbonias 1999). It is usually assumed that small and diffuse artifact concentrations escape detection more easily than large and dense scatters of material (Flannery 1976; Cherry 1983; Barker 1995a). To balance this effect, in this study small sites such as Hellenistic farms were treated separately from the other settlement categories (*e.g.* sites interpreted as villages or villas). Site categories were formulated for each site by the LERC survey team. Therefore, we could easily make a selection of small Hellenistic sites ('farms'). In this analysis, we considered both the totality of sites recorded (independently from period and category) (99), the Hellenistic settlements (81) and the small Hellenistic farms (62). More precisely, in order to assess the role played by ground visibility in site discovery, we controlled the surface conditions characterizing those units where the actual discovery of sites took place. If a site extended over more units, only the visibility conditions of the first unit walked were considered in this analysis since this is where the discovery, in the first place, happened.

4.4.1 PROCEDURE

Technically speaking, significant associations (if any) between the proportions of discovered sites and survey visibility conditions are evaluated through nonparametric one-sample tests (for a good overview of one- and two-sample tests see Siegel 1956, 35-

156; Shennan 1988, 57, 104-126; Kvamme 1990; Wheatley & Gillings 2002, 123-132; Conolly & Lake 2006, 112-148). These tests are of the goodness-of-fit type (Siegel 1956, 35): they compare the observed sample with a theoretical distribution in order to single out unexpected anomalies in the frequency of the phenomenon under consideration (in our case, site discovery).

For the visibility factors which are nominal in type (*i.e.* composed of different categories, for example land use) we applied the chi-square test to see whether a significant difference existed between the observed and the expected proportions of sites. For testing the influence of the ordinal variables (*i.e.* variables for which values can be ordered on a ranking scale such as vegetation cover rate, stoniness, etc.) we preferred instead the Kolmogorov-Smirnov test, which measured the strength of the divergence between the observed and the expected cumulative frequency distributions (Siegel 1956, 47-52; Shennan 1988, 53-61).

In addition to these approaches, we also used the analytical technique devised by Attwell and Fletcher (1985; 1987) to overcome some intrinsic limitations of the previous two tests. This test of association, unlike the other two, has the advantage of indicating both the magnitude and the direction (positive or negative) of the association in each category and can be used on small samples or categories. We relied especially on the results of the Attwell-Fletcher technique because it is more sensitive and powerful than the other two tests (Attwell & Fletcher 1985; 1987; see also the discussion in Siegel 1956, 46 - 52).

As we will see, the Attwell-Fletcher test indicated a significant association only with the tillage variable. However, because other factors, such as the presence of overgrown vegetation, are known to prevent the surveyors from properly seeing the ground while surveying (*e.g.* Terrenato & Ammerman 1996; Terrenato 2000), we also considered the other factors in the following Multi Criteria Evaluation. It is worth remembering that our aim here was to visualize those landscape locations where visibility conditions were optimal for site discovery independently of whether or not a site was actually found. This allowed us to pinpoint reliable settlement vacuums in the Hellenistic pattern (but see the discussion below on geomorphological biases).

Before proceeding with the Multi Criteria Evaluation, we wanted to assess a possible linear relationship (collinearity) between the explanatory variables (visibility factors) (Shennan 1988, 177-179; Vaughn & Crawford 2009; Langston 2013, 62, 112). Collinear variables are dependent variables, which means that they depend on each other and are related to the very same phenomenon. These factors are redundant predictors and thus some can be excluded from the model. For instance, it is very likely that the degree of shadow could be highly correlated with other factors that contribute to the shading of the field, such as the vegetation cover rate (which is directly proportional to the shadow) or the soil humidity (which, like the shadow factor, depends on the time of survey and on the weather conditions).

The visibility factors were therefore tested for collinearity by means of the Band collection statistic tool of ArcGIS 10.2.2 (Esri 2014). The threshold of 0.6 was chosen for the exclusion (Langston 2013, 116). Tillage and land use did not exhibit collinearity and were selected as suitable independent predictors. On the other hand, stoniness, soil humidity, shadow condition, recent material, and vegetation cover rate exhibited high collinearity (index of correlation > 0.8). As mentioned earlier, using them simultaneously would have entailed redundancy in the model. Therefore, only one of them was selected as candidate predictor for the calculation of the predictive map of visibility. We opted for the vegetation cover factor for two main reasons: among these collinear factors it is the most tangible and the most objectively definable by surveyors in the field.

These three independent visibility factors were then reclassified in visibility scores according to experts' judgments, in order to obtain criterion maps for the MCE tool. The procedure of using experts' judgments for assigning scores and weights of importance is a common practice in predictive modeling (Judge & Sebastian 1988; Van Leusen & Kamermans 2005; Verhagen 2007). The team leaders that had been working in the various LERC campaigns in

Isernia gained quite some experience in surveying this landscape and thus were the most appropriate persons for assigning visibility scores from one (low) to three (high) to the categories of the three selected visibility factors (tillage, land use, vegetation cover rate) (Table 4.1).

Afterwards, weights of importance were appointed to these three variables according to the statistical results (Tables 4.2 – 4.4) and the experts' personal judgments. These weights numerically represent the different influence each factor theoretically would play in site discovery. According to the survey team leaders and the statistics, tillage plays the most significant role in influencing the discovery of sites, followed by land use and then by vegetation cover rate. The Pairwise technique of IDRISI GIS – Selva edition (Eastman 2012, 133-134) was applied to carry out this weight calculation and the following rank of weights was established (consistency 0.03, the sum of the weights must be 1):

1) Tillage: 0.7514
2) Land use: 0.1782
3) Vegetation cover rate: 0.0704

Once the visibility factors were reclassified in criterion maps, they could be merged through the MCE tool to finally obtain a predictive map of the suitability for site discovery based on survey visibility conditions (regardless of ancient settlement strategies or recent geomorphological erosive or depositional biases). The three criterion maps representing respectively tillage, land use and vegetation rate were combined in a Weighted Linear Combination (WLC) according to the following formula (Eastman 2012, 132):

$$S = \sum w_i x_i$$

S = suitability for site discovery based on visibility factors, w_i = weight of visibility factor i and x_i = criterion score of visibility factor i.

4.4.2 RESULTS

The Attwell-Fletcher results showed that at a 95% level of probability the finely plowed units (tillage factor) seem to be favorable for the discovery of all sites, Hellenistic settlements, and small Hellenistic farms (Tables 4.2 – 4.4). In spite of the fact that only 3.3% of the area surveyed by the LERC team was finely plowed, the number of archaeological sites recorded there is significant (more sites than expected from a random distribution). The Attwell-Fletcher test did not identify any other significant association for site discovery with the other visibility factors, neither for the totality of sites nor for the Hellenistic settlements and small farms.

Table 4.1 Experts' visibility scores for the variables used in the MCE analysis.

Tillage	**Visibility score: from 1 (low) to 3 (high)**
none	1
rolled	3
harrowed	3
finely plowed	3
medium plowed	2
heavily plowed	2
Land use	
Arable / arable cleared	3
Arable completely cultivated	1
Horticulture	1
Olives / fruit trees	1
Viticulture	2
Fallow	2
Fallow clean / burnt	2
Fallow with stubs	2
Wood / *macchia* / fallow overgrown / pasture	1
Vegetation cover rate	
0	3
1	3
2	2
3	2
4	1
5	1

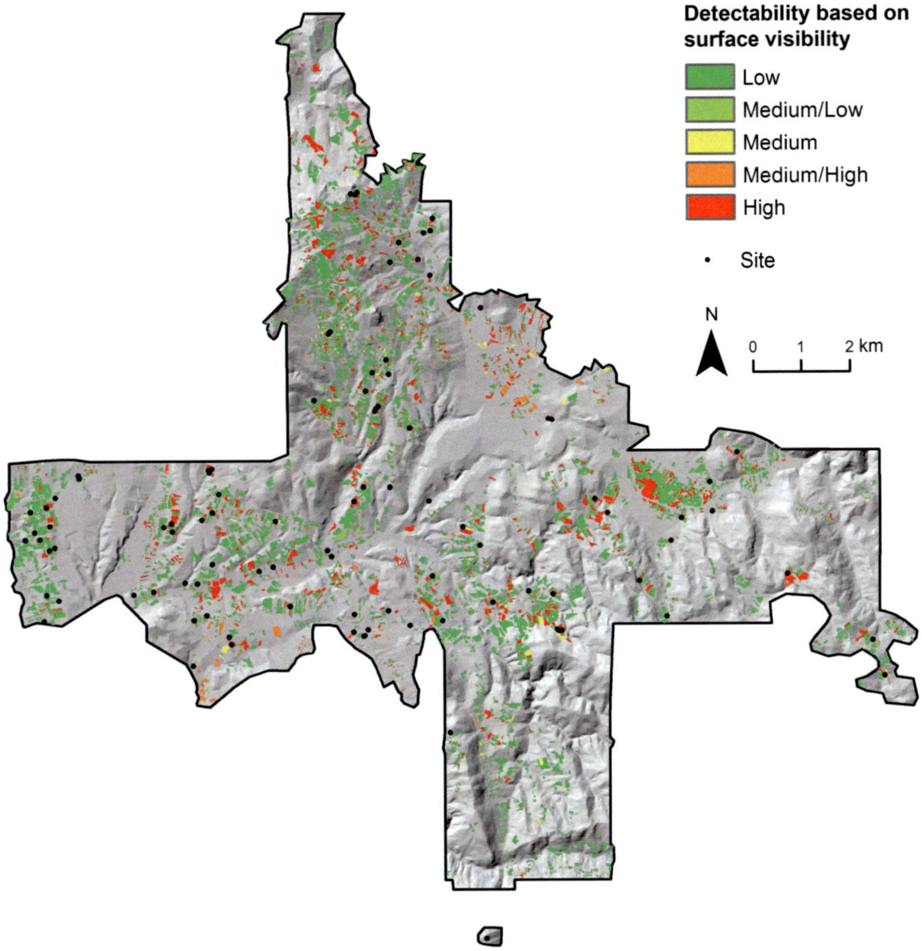

Fig. 4.2 Predictive map of the suitability for site discovery in the surveyed units based on surface visibility conditions (*i.e.* detectability map based on surface visibility). The probability for site discovery (detectability) based on surface visibility goes from low to high and is indicated in each unit by the different colors of the palette. Raster base map: shaded relief calculated from the 10 m-resolution DEM named TINITALY/01 (Tarquini *et al.* 2007; 2012; Tarquini & Nannipieri 2017).

As a last step of this first analysis, a value of suitability for site discovery was assigned through the Multi Criteria Evaluation (MCE) to each surveyed unit, which is the result of the sum of the visibility factor scores (*i.e.* the scores for tillage, land use, and vegetation cover rate indicated in Table 4.1), characterizing that unit, multiplied by their respective factor weights (see before). Such a predictive map (Fig. 4.2) highlights effectively reliable empty spaces in the recorded settlement pattern or, in reverse, possibly unreliable vacuums where there is the risk of missing some archaeological evidences due to the critical survey visibility conditions.

As has been demonstrated, visibility factors may favor or prevent the discovery of sites (*e.g.* Terrenato & Ammerman 1996). However, in our case, considering the widely scattered configuration of the few surveyed units characterized by finely plowed surfaces, the previously detected association with the tillage factor is insufficient for explaining the entire regional pattern recorded. Other factors may have affected the overall pattern much more significantly, such as large-scale gradational processes of erosion and deposition. We turn our attention now precisely to these processes.

Table 4.2 Attwell-Fletcher test for the tillage variable. Sample: totality of sites (99). Surveyed area 1886.031 ha. Number of simulations: 200. Critical values: 95th percentile = 0.43 ± 0.016; 5th percentile = 0.00 ± 0.000. Notes: the Kolmogorov-Smirnov test indicates a maximum difference of 0.283 between the observed and the expected cumulative frequency distributions (critical value to reject the null hypothesis of no association is 0.137, with α = 0.05).

Tillage class	Area (ha)	Number of sites	Expected proportion of sites	Observed proportion of sites	Category weight	Positive association	Negative association
None	1426.880	47	0.757	0.47	0.06	no	no
Rolled	22.054	1	0.012	0.01	0.08	no	no
Harrowed	51.488	5	0.027	0.05	0.17	no	no
Finely plowed	62.601	18	0.033	0.18	0.50	yes	no
Medium plowed	276.398	27	0.147	0.27	0.17	no	no
Heavily plowed	46.609	1	0.025	0.01	0.04	no	no

Table 4.3 Attwell-Fletcher test for the tillage variable. Sample: Hellenistic settlements (81). Surveyed area 1886.031 ha. Number of simulations: 200. Critical values: 95th percentile = 0.46 ± 0.022; 5th percentile = 0.00 ± 0.000. Notes: the Kolmogorov-Smirnov test indicates a maximum difference of 0.238 between the observed and the expected cumulative frequency distributions (critical value to reject the null hypothesis of no association is 0.151, with α = 0.05).

Tillage class	Area (ha)	Number of Hellenic settlements	Expected proportion of settlements	Observed proportion of settlements	Category weight	Positive association	Negative association
None	1426.880	42	0.757	0.52	0.06	no	no
Rolled	22.054	1	0.012	0.01	0.09	no	no
Harrowed	51.488	5	0.027	0.06	0.20	no	no
Finely plowed	62.601	15	0.033	0.19	0.49	yes	no
Medium plowed	276.398	17	0.147	0.21	0.12	no	no
Heavily plowed	46.609	1	0.025	0.01	0.04	no	no

Table 4.4 Attwell-Fletcher test for the tillage variable. Sample: Hellenistic farms (62). Surveyed area 1886.031 ha. Number of simulations: 200. Critical values: 95th percentile = 0.47 ± 0.021; 5th percentile = 0.00 ± 0.000. Notes: the Kolmogorov-Smirnov test indicates a maximum difference of 0.305 between the observed and the expected cumulative frequency distributions (critical value to reject the null hypothesis of no association is 0.172, with α = 0.05).

Tillage class	Area (ha)	Number of Hellenic farms	Expected proportion of farms	Observed proportion of farms	Category weight	Positive association	Negative association
None	1426.880	28	0.757	0.45	0.04	no	no
Rolled	22.054	1	0.012	0.02	0.10	no	no
Harrowed	51.488	4	0.027	0.06	0.18	no	no
Finely plowed	62.601	14	0.033	0.23	0.51	yes	no
Medium plowed	276.398	14	0.147	0.23	0.12	no	no
Heavily plowed	46.609	1	0.025	0.02	0.05	no	no

4.5 EVALUATING GEOMORPHOLOGICAL BIASES

The aim of our next analysis is to assess the extent to which recent erosion and deposition may affect the preservation of the Hellenistic-period archaeological record in the Isernia basin (Coltorti 1983; Van Otterloo & Sevink 1983; Stek *et al.* 2015). Post-depositional geomorphological processes that have been occurring over the Late Pleistocene and Holocene may be a source of bias for the distribution of Prehistoric and later sites. For the period of interest to us (*i.e.* the Hellenistic period) we focused on the most recent Late Holocene features and gradational processes (*i.e.* 2500 B.C. - present) in order to point out critical zones where deposition or erosion may have covered or deleted Hellenistic sites after their abandonment.

Gradational processes comprise both degradational and aggradational processes (Bos & Sevink 1975). The former stand for erosive forces acting in a landscape to destroy, remove, and occasionally also uncover (Ebert & Singer 2004) the archaeological remains after the primary deposition (*i.e.* in our case the Hellenistic period). The latter indicate the processes leading to sedimentary accumulative *facies* (alluvium and colluvium) and may result in the burial of sites (see the discussion in Cremaschi & Nicosia 2012).

Soil formation requires landscape stability in order to let the rate of pedogenesis exceed that of erosion or deposition (Bos & Sevink 1975, 223). In other words, when erosion or accumulation on previous unstable surfaces stops, pedogenesis can start and, eventually, a soil is formed on these now stable land surfaces, whether in the sediment accumulated (following aggradation) or the truncated land surface (following degradation), with eventually remainders of the former soil. A given soil thus represents the record of the gradational history of the land surface concerned, allowing for making inferences about its pedogenesis and formative history.

4.5.1 PROCEDURE

We used the information provided by the descriptions of the soil types by van Otterloo (1981; Stek *et al.* 2015, 291) and Koopmans (1980), to assign, within the survey sample area, gradational scores to each of the individual mapping units of the integrated soil map (1: 25,000, see Fig. 4.3 p. 79) produced by van Otterloo and Sevink. Scores range from 0 (recent gradational processes almost absent) to 3 (recent gradational processes very prominent). In this way, the landscape stability of each soil map unit was classified. On the basis of this evaluation we then established the detectability of Hellenistic sites, that is the probability of finding such sites during field survey, regardless of the influence of ancient location preferences (see Appendix A). However, the scale and the temporal resolution of the soil map allowed us to make an evaluation of the depositional and erosive processes only at a regional scale (1:25,000), and we did not reach more detailed assessments. For instance, due to the relatively large scale of the soil map, we could not evaluate the effect of possible smaller-scale geomorphological phenomena observable only at a finer spatio-temporal resolution such as at the unit-scale resolution represented by modern farming fields (see the discussion in Butzer 2008, 403 - 404). For example, it is not possible to assess through this soil map the occurrence of potential local transport of topsoil related to modern mechanized agricultural activities performed within the field, such as plowing or land-leveling for field clearance and preparation.

While it is demonstrated that such small-scale processes could affect the distribution of artifacts in the plough-soil (*e.g.* Given 2004, 18 - 19), it remains unclear to what extent these local processes affect large-scale patterns such as regional distributions of archaeological sites (but see the case-study in Diez-Martín 2010). With regard to the Aesernia case-study, the difference in scale between the mapping of the field units and the soil units might have influenced in certain zones of the landscape our analysis of the reliability of our regional pattern of sites. As long as higher resolution soil maps are not available, the only way to assess the potential impact of small-scale geomorphological processes on the position and/or extension of sites detected at the surface, would be through small-scale targeted excavations or probing. Such ground truthing campaigns aimed at testing our predictions on the date and character of colluvial and erosive processes, and their effects on site presence and position, are scheduled for the future.

In this analysis, however, we concentrate only on the effects that recent large-scale erosive or accumulative phenomena may have had on the visibility and connected discovery of archaeological sites. The diagnostic indicators we used can help disentangle degradation (erosion) and aggradation (accumulation) processes. Degradation was evaluated through judgment of the soil profile development, particularly the soil horizon differentiation, and eventual extent of truncation of the soil concerned. For instance, well-developed and well-drained soils on old stable surfaces, exhibiting all the characteristics of a non-truncated and well-preserved soil, notably a well-developed A horizon and eventually E horizon, indicate that natural or anthropogenic erosion (plow erosion) and colluviation did not occur over a very prolonged period of time. On the contrary, soils that exhibit a pronounced argic horizon, but lack an eluvial horizon and have a poorly developed Ah horizon, evidently have been recently truncated, losing their topsoil.

An example of a unit with well-preserved soil is unit 10 (see Figs. 4.3 – 4.4 and Appendix A). The Eutric Nitosol (FAO/UNESCO 1974) on fluvio-lacustrine deposits is a well-developed polygenetic soil of old, stable, and sub-horizontal surfaces where only minor recent erosion occurred, evidenced by the presence of well-developed A and E horizons. For these reasons, we established that the detectability of Hellenistic sites at this unit was high.

Evidently, land surfaces may also be so recent that they postdate the Hellenistic period and in that case chances for finding Hellenistic sites do not exist or are very low. This is the case when serious degradation took place after that period, and may be the situation on truly unstable land surfaces, *e.g.* steep, unstable slopes. Chances for such sub-recent strong degradation can be judged by evaluation of the extent of soil formation. An example are units 32 - 34 (see Figs. 4.3 - 4.4 and Appendix A), in which Rendzinas and Regosols dominate on relatively unstable and steep marls and shales, with prominent active erosion. The limited extent of soil formation and currently active erosion strongly suggested that the current land surface is of a younger age and thus the detectability of Hellenistic sites was low.

As regards the evaluation of land surfaces that exhibit aggradation, criteria very much resemble those described above: crucial is the extent of soil formation and soil horizon differentiation, which provides a clear indicator for the time elapsed since the deposition of the sediment in which the soil has formed. In fact, this approach forms the basis for

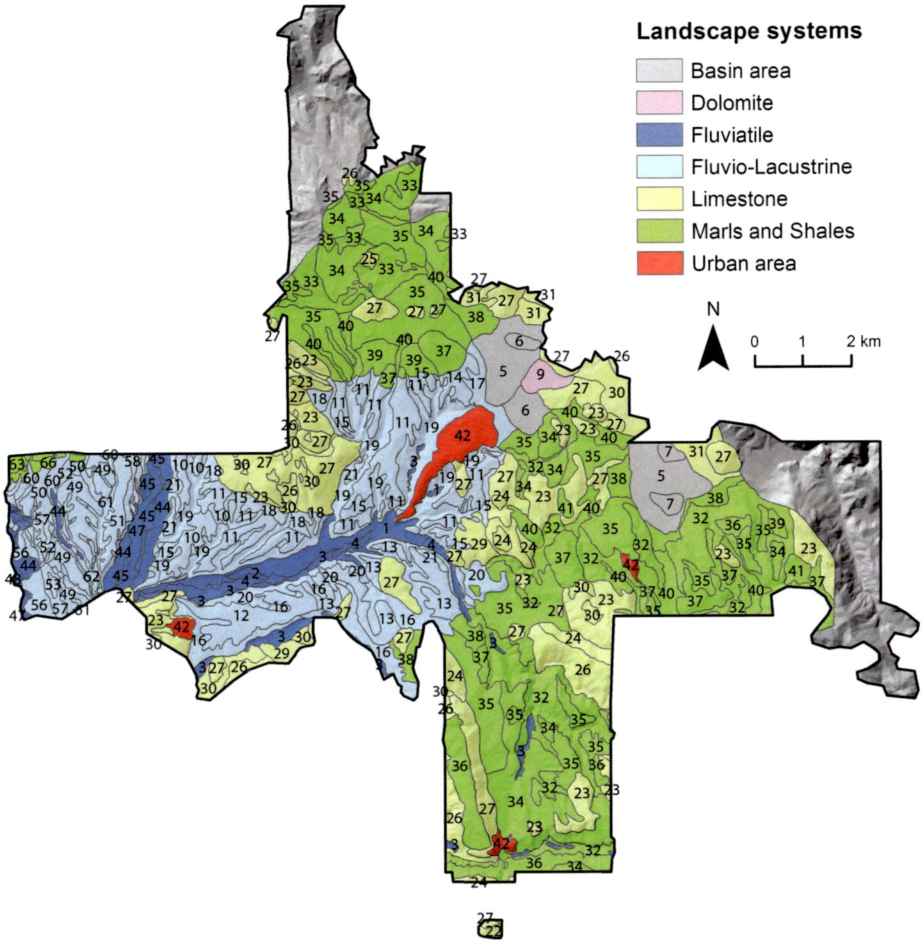

Fig. 4.3 Soil map, 1: 25,000 by van Otterloo and Sevink (van Otterloo & Sevink 2016). The numeric labels indicate the soil units (see Appendix A). Raster base map: shaded relief calculated from the 10 m-resolution DEM named TINITALY/01 (Tarquini *et al.* 2007; 2012; Tarquini & Nannipieri 2017).

all studies on soil chronosequences and has led to considerable insight into the rate of soil formation as dependent on climate and parent material (see for example Sevink *et al.* 1982; Sauer *et al.* 2012). Evidently, also other criteria were used by van Otterloo and Koopmans in their assessment of the age of sediments encountered, in particular the presence of archaeological material in these sediments. Such a criterion has been extensively used in soil studies, paying attention to the age and origin of anthropogenic colluvial topsoils in the Mediterranean (*e.g.* Remmelzwaal 1978; Lang & Bork 2006).

There is another, very region-specific, diagnostic indicator in the evaluation of these soil map units. Throughout the Isernia area, the latest major Holocene tephra layers, called the Avellino pumice layer (near 2000 B.C.) and the Vesuvius Pompeii ash layer (79 A.D.) have been deposited and thus should be present in "stable" soils. They were easily identified through the dark color of the topsoil, because rather than distinct layers the volcanic material often appeared as abundant fine angular pyroclastic particles. Therefore, the presence in the topsoil of pyroclastic material was a proxy in aid of the evaluation whether a soil has been recently truncated or 'aggraded' after the last eruptions (for a discussion on ash fall as an excellent temporal marker for studying gradational processes in the Mediterranean see Judson 1963; Vita-Finzi & Judson 1964; Brown 1997, 239-242; Lefèvre *et al.*

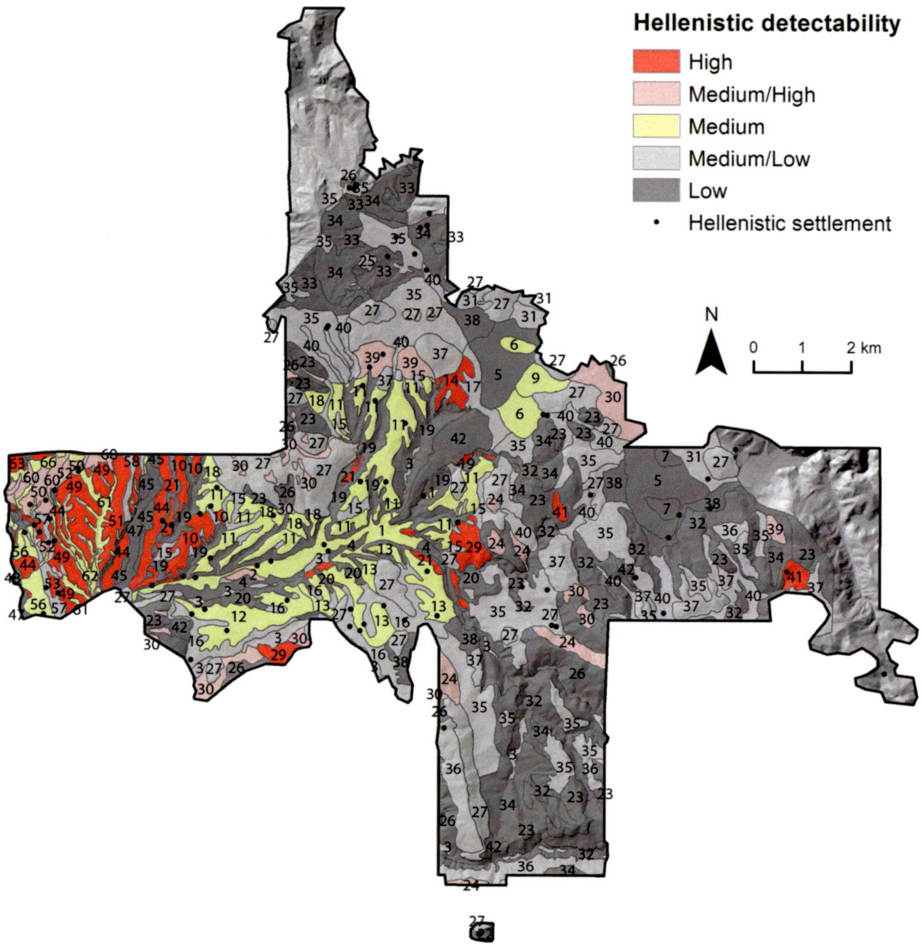

Fig. 4.4 Detectability map for the Hellenistic period based on geomorphological processes (*i.e.* geomorphological detectability map). Raster base map: shaded relief calculated from the 10 m-resolution DEM named TINITALY/01 (Tarquini *et al.* 2007; 2012; Tarquini & Nannipieri 2017).

2010). If not present, we could conclude that soils were formed on late Roman or younger deposits, or on strongly truncated, more recent land surfaces and thus are devoid of Hellenistic and earlier sites. For an extensive review of the Late Pleistocene and Holocene ash falls in the Isernia area, reference is made to Stek *et al.* (2015, 241-276).

In areas where such unique ash falls did not occur, the dating of soils, their truncation, and the age of the deposits in which they were formed in case of relatively young soils, is more problematic to establish and often far less reliable, having to be based on rather tentatively established rates of soil genesis during early soil formation, *e.g.* decalcification and accumulation of organic matter (*e.g.* Cremaschi 1987). Both processes are very much climate and parent material dependent, and do not allow for more than broad statements on soil age. Processes like clay translocation and weathering were not discriminative at the time scales concerned, and were thus unsuited for our purposes (see *e.g.* Sevink *et al.* 1982; Cremaschi & Sevink 1987; Sauer 2010).

An example is provided by the Le Piane karst basin located northeast of Isernia (unit 5, see Figs. 4.3 – 4.4 and Appendix A). Here post-Roman colluvial fill deposits, devoid of pyroclastic material, cover any previous archaeological remains. These colluvia probably originated through strong erosion impacting on the unstable surfaces of the surrounding mountains, and subsequent deposition of the eroded soil material in this basin, which may be related to intensive anthropogenic deforestation and agricultural activities especially in Roman times (see the discussion in Burri *et al.* 1999).

4.5.2 RESULTS

Through the evaluation of the soil maps we could identify and delineate zones where erosion and sedimentation likely took place recently to various extents (Fig. 4.4, p. 80). In archaeological surveys, these critical units (in dark grey in Fig. 4.4) can be expected to more likely exhibit a biased archaeological record: the higher risk of Late Holocene erosion and deposition at these units must be taken into account when evaluating the reliability of the recorded Hellenistic archaeological evidence.

On the basis of Fig. 4.4, a potential bias possibly hampering the detection of sites seems strong in the south transect, and to a lesser extent in the central, east, and north parts of the survey sample area, whereas in the west transect its effect should be minimal in light of the predominantly highly stable surfaces. We used a statistical test to assess the probability that the archaeological pattern is indeed influenced by this factor.

Statistical tests of association can systematically compare the proportion of the total sites found by the LERC team in a certain detectability class to the proportion of the total survey coverage at that detectability class (*i.e.* the proportion represented by the area walked in that class, see Fig. 4.5), and tell us whether or not this association is statistically significant. In order to assess whether a significant association exists between detectability class and site discovery the Attwell-Fletcher statistical test of association was performed within the surveyed units (Attwell & Fletcher 1985; 1987). This test indicated, for the surveyed units in each detectability class, whether there were significantly more, or significantly fewer sites than expected *vis-à-vis* random distributions. These random distributions theoretically represent site distributions which configurations are independent from the variable under consideration (*i.e.* geomorphological detectability): according to the extent of the area surveyed in each detectability class, sites were thus proportionally allocated by means of several simulations (in our case 200 simulations were run). This means that the site proportions of these simulated random distributions that were allocated in the various detectability classes (*i.e.* expected proportions of sites) corresponded to the proportions of the total surveyed area at the detectability classes. If a significant negative association existed (*i.e.* in that class, there were fewer observed sites than expected from distributions unaffected by geomorphological processes), the category weight of the detectability class under consideration resulted lower than the critical value for the 5th percentile. Conversely, if there was a significant positive association (*i.e.* more recorded sites than expected), the category weight exceeded the critical value for the 95th percentile (see Tables 4.5 – 4.6).

As shown in Tables 4.5 – 4.6 there seems to be a positive association with the medium/high detectability class (covering 7.1% of the total surveyed area, *i.e.* a proportion of 0.071 : 1) and the presence of Hellenistic settlements and farms. This may be explained by the relatively stable surfaces characterizing this land class, which are favorable for the preservation of Hellenistic settlement sites (see for instance the Valle Porcina in the West transect, Fig. 4.1). However, this does not completely exclude the possibility that, rather than more preserving landscape units, ancient location preferences instead caused sites to cluster in certain zones. This hypothesis is supported by the fact that we did not discern any other significant correlation between detectability classes and site numbers: indeed, site numbers in the theoretically low and medium/low detectability zones were actually not significantly lower than expected. More importantly perhaps, there was also no positive correlation between site numbers and the high detectability zones. Despite the highly stable surfaces associated with these zones, they did not show significantly higher percentages of sites.

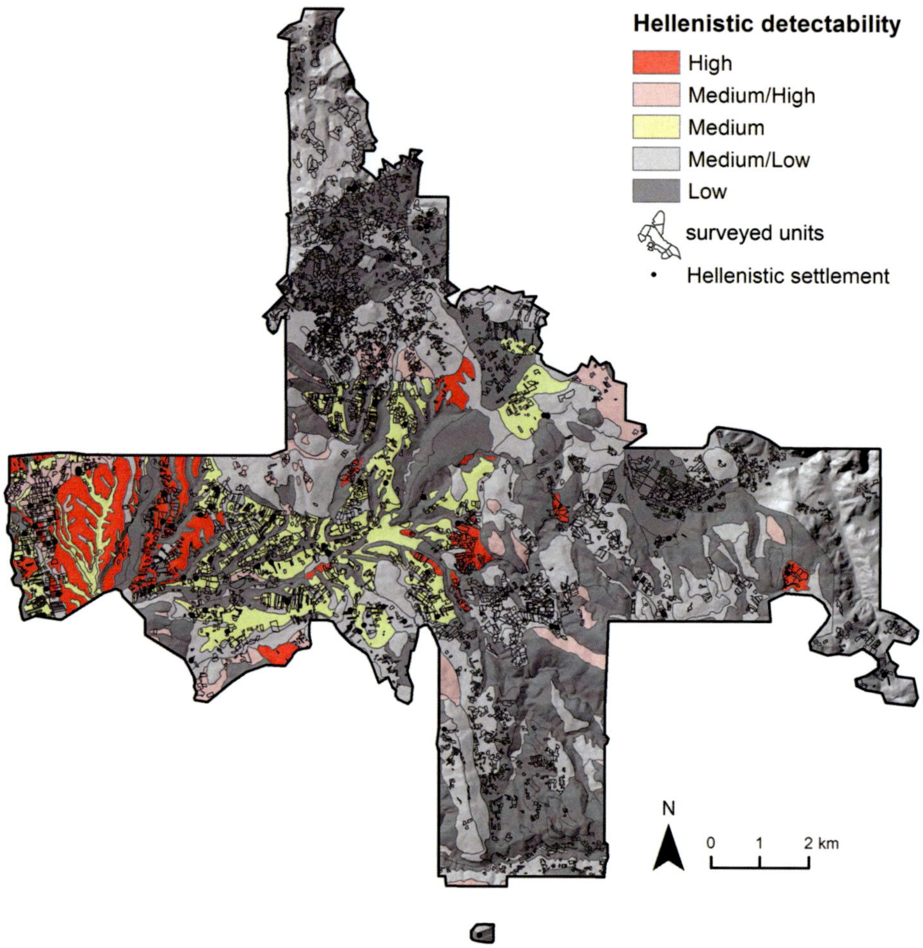

Fig. 4.5 Geomorphological detectability map for the Hellenistic period with the indication of the surveyed units within each detectability unit. Raster base map: shaded relief calculated from the 10 m-resolution DEM named TINITALY/01 (Tarquini *et al.* 2007; 2012; Tarquini & Nannipieri 2017).

Therefore, the overall picture that emerged from this analysis is that there is not always a clear correlation between site recovery rates and detectability classes in the investigated territory of Isernia. As a matter of fact, if geomorphological detectability was responsible to a significant degree for the discerned regional settlement pattern, we would have expected to find positive correlations with the high and medium/high classes and also negative correlations with the medium/low and low classes. Consequently, the low site density areas recorded in several large portions of this territory, and the few localized-higher density zones widely scattered (Fig. 4.1), could plausibly be the result of genuine Hellenistic settlement patterns. This hypothesis is supported especially by the low site numbers recorded for the high detectability units. Considering the minimal susceptibility to post-Hellenistic-period erosion and deposition, in these units, it seems implausible to assume that the recorded pattern of thinly populated zones in the Hellenistic period is entirely the result of geomorphological biases.

The fact that there were not significantly fewer sites than expected in the low detectability conditions may possibly be related to the large sample of surveyed units in the medium/low and low detectability zones affected by high erosion (erosion score 2 or 3, see Appendix A). It is

Table 4.5 Attwell-Fletcher test of association on the Hellenistic settlements. Total number of settlements within the geomorphological detectability map: 77. Surveyed area within the geomorphological detectability map: 1688.93 ha. Number of simulations: 200. Critical values: 95th percentile = 0.34 ± 0.012; 5th percentile = 0.04 ± 0.022.

Geomorphological detectability class	Area (ha)	Number of Hellenistic settlements	Expected proportion of settlements	Observed proportion of settlements	Category weight	Positive association	Negative association
High	110.382	6	0.065	0.08	0.19	no	no
Medium/High	119.928	12	0.071	0.16	0.36	yes	no
Medium	375.959	25	0.223	0.32	0.24	no	no
Medium/Low	578.492	26	0.343	0.34	0.16	no	no
Low	504.172	8	0.299	0.10	0.06	no	no

Table 4.6 Attwell-Fletcher test of association on the Hellenistic farms. Total number of farms within the geomorphological detectability map: 58. Surveyed area within the geomorphological detectability map: 1688.93 ha. Number of simulations: 200. Critical values: 95th percentile = 0.37 ± 0.013; 5th percentile = 0.00 ± 0.025.

Geomorphological detectability class	Area (ha)	Number of Hellenistic farms	Expected proportion of farms	Observed proportion of farms	Category weight	Positive association	Negative association
High	110.382	5	0.065	0.09	0.21	no	no
Medium/High	119.928	10	0.071	0.17	0.38	yes	no
Medium	375.959	15	0.223	0.26	0.18	no	no
Medium/Low	578.492	22	0.343	0.38	0.17	no	no
Low	504.172	6	0.299	0.10	0.05	no	no

important to underline that erosive agents can destroy sites, but they can also uncover otherwise-buried and thus previously-invisible archaeological material (see Ebert, & Singer, 2004). Field-walkers often find sites at places where erosion is active, and this may be also the case for some of the sites found in our study area. For example, where Late Holocene degradational processes are prominent and still active (*e.g.* low detectability units 32 – 34) sites might have been occasionally exposed at the surface by forces of erosion (*e.g.* local slope movements, plowing).

Overall, if we accept the validity of the principle used here to predict the effect of large-scale geomorphological biases (in other words, strong recent deposition may cover Hellenistic sites, and strong recent erosion may destroy or occasionally uncover Hellenistic sites), on the basis of our tests we may therefore conclude that this analysis underscores the value of the Aesernia survey data for studying past settlement behavior at a regional scale of analysis.

4.6 COMBINING RESULTS

By means of a cross-tabulation, the previous two detectability maps, based on surface visibility and geomorphology respectively, were combined in order to assign to each surveyed unit a pair of detectability scores which aid archaeologists to evaluate the combined effect of these two sets of biasing factors.

As shown in Table 4.7 and in Fig. 4.6, the combination of these two types of information allows the archaeologist to assess for each surveyed unit whether the density and pattern of sites recorded

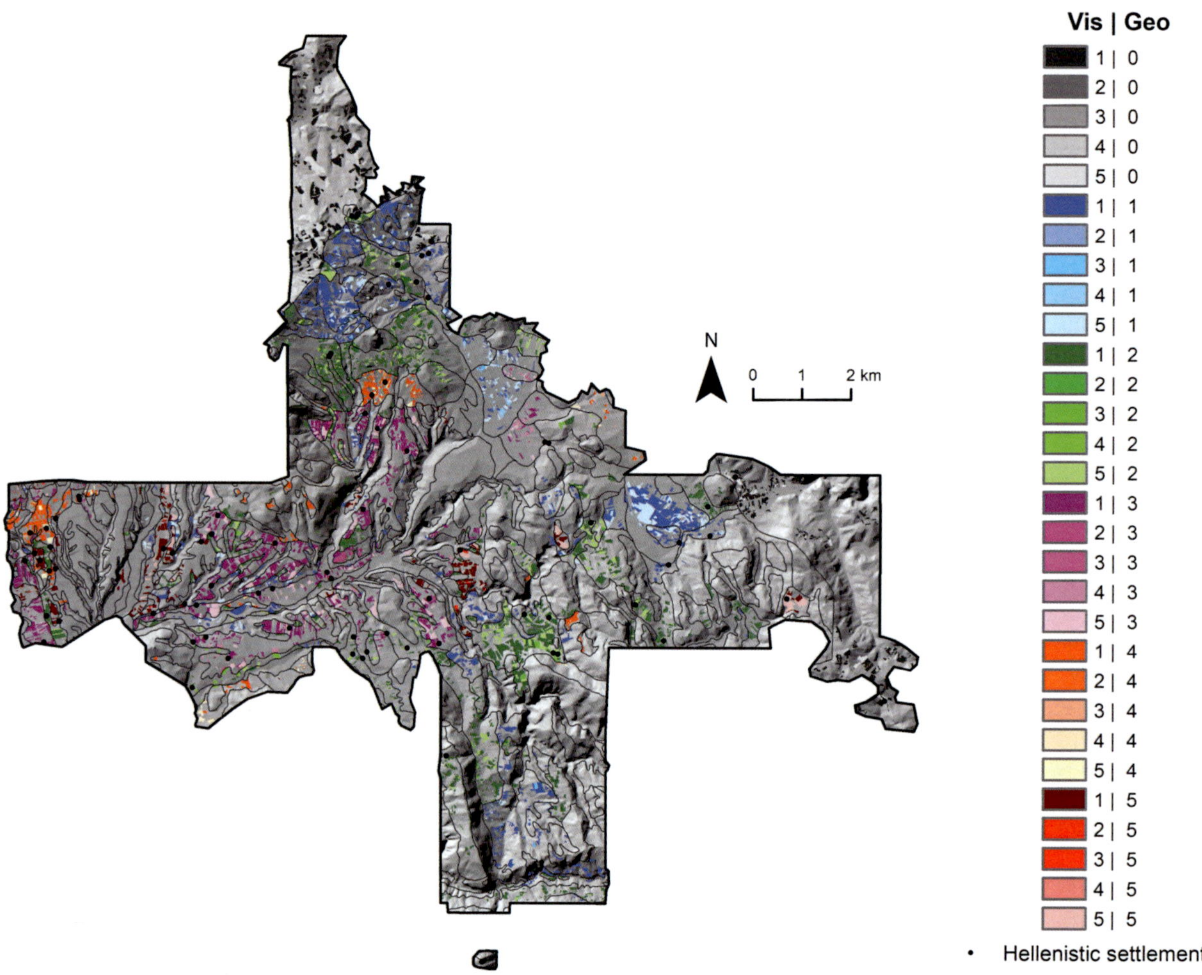

Fig. 4.6 Cross-tabulation between the detectability map based on surface visibility and the geomorphological detectability map. Each surveyed unit receives a combination of detectability scores (the first score refers to the detectability based on surface visibility, the second score refers to the Hellenistic detectability based on the geomorphological evaluation): 0 = outside soil map coverage; 1 = low; 2 = medium/low; 3 = medium; 4 = medium/high; 5 = high. Raster base map: shaded relief calculated from the 10 m-resolution DEM named TINITALY/01 (Tarquini *et al.* 2007; 2012; Tarquini & Nannipieri 2017).

Table 4.7 Cross-tabulation between surface visibility classes (columns) of the predictive surface visibility map (see Fig. 4.2, p. 75) and geomorphological detectability classes (rows) of the geomorphological detectability map (see Fig. 4.5, p. 82). The numbers indicate the percentage (%) of the surveyed territory (1886.031 ha) for each combination of classes.

Geomorphological detectability	**Detectability based on surface visibility**					
	Low	Medium/Low	Medium	Medium/High	High	Total
outside soil maps	8.18	0.03	0.12	0.04	2.25	10.62
Low	20.16	0.29	0.33	0.33	5.62	26.74
Medium/Low	22.98	0.19	0.89	0.48	6.15	30.7
Medium	14.25	0.38	0.36	0.72	3.96	19.67
Medium/High	5.07	0.06	0.23	0.3	0.71	6.37
High	4.05	0.08	0.13	0.06	1.58	5.9
Total	74.69	1.04	2.06	1.94	20.27	100

may correlate with either, or both, types of bias. The complexity of the interrelationships between surface visibility and geomorphology can be better elucidated with two examples. First, a unit may display optimal surface visibility conditions, but despite a maximum visibility score Hellenistic evidence will hardly be found if covered under post-Hellenistic deposition. Second, stable-old surfaces may have a high detectability score, but thick vegetation may still prevent archaeologists from detecting a site.

Surface visibility and geomorphological biases are intricately interwoven, and their complex relationships are difficult to predict without a suitable support for systematic consultation: if we aim to visualize the potential implications of these relationships for the survey record, we need to find a means to calculate them and then represent them simultaneously on a map. A combined detectability map as the one displayed in Fig. 4.6 may fulfill this task. Archaeologists may use this map as a filter to be superimposed on a site distribution map compiled during field survey. In this way, they may eventually be able to filter out potential distortions provoked by visibility and geomorphological processes on site numbers and pattern, calibrate the pattern, and possibly correct it through the simulation of the potentially missing evidence and then, finally, assess differences in the historical interpretations they will put forward before and after such a calibration.

4.7 CONCLUSIONS

By definition survey data are fragmentary. Nonetheless, archaeologists by necessity have to base their reconstructions on here-and-there surfaces where preserved and uncovered archaeological material can be observed. When investigating regional settlement patterns and site numbers as reflected by field survey data, it is thus crucial to understand the formative processes leading to the creation of the field survey record. Assessing the constraints that may affect the preservation and the recording of sites in different portions of the landscape should be a first step to take before inferring ancient location preferences, land-use strategies and, more broadly, historical processes. Detectability maps can help achieving this goal.

Not only are detectability maps useful to test for biases in survey data, they can also help designing a targeted sampling strategy for future surveys. In this way, detectability maps can increase the efficiency and reliability of survey methods and data: stable and well-visible surfaces, of course, are more likely to return a more representative settlement density and pattern than those where erosion and sedimentation are, or have been recently, occurring.

Detectability maps are by no means a new tool. However, the innovative aspect of our method is that it offers a formal, and widely applicable, procedure for combining a systematic statistical analysis of visibility biases with a straightforward assessment of regional geomorphological biases based on soil characteristics. The resulting maps that highlight the position of visible and less visible, stable and unstable surveyed surfaces provide a filter through which archaeologists can calibrate or re-evaluate interpretations of recorded site distributions. With its formal character, the method proposed here may also contribute to overcoming current challenges in the comparison and integration of large-scale survey data on the supra-regional level.

As regards the case of the colony of Aesernia, we used the detectability maps to assess whether the few localized high site densities and the large tracts of empty zones in between them may be the result of visibility and geomorphological biases, or rather the result of settlement location preferences of ancient communities. We observed that low site density and blank areas are also frequently present in areas characterized by good surface visibility and high geomorphological detectability. This conclusion strengthens the reliability of the recorded empty units in between clusters of sites and, thus, further supports the clustered early colonial settlement hypothesis recently proposed for this Apennine landscape in the Hellenistic period.

4.8 REFERENCES

Allen, M. J. 1991. Analysing the landscape: A geographical approach to archaeological problems. In A. J. Schofield (ed.), *Interpreting artefact scatters: Contributions to ploughzone archaeology*, 39-57. Oxford: Oxbow Monograph 4.

Arnoldus-Huyzendveld, A. 2007. Le trasformazioni dell'ambiente naturale della pianura grossetana. Dati geologici e paleo-ambientali. In C. Citter & A. Arnoldus-Huyzendveld (eds.), *Archeologia urbana a Grosseto. Origine e sviluppo di una città medievale nella "Toscana delle città deboli". Le ricerche 1997 – 2005. I: La città di Grosseto nel contesto geografico della bassa valle dell'Ombrone*, 62-71. Firenze: All'Insegna del Giglio.

Arnoldus-Huyzendveld, A. 2011. Landscape development of the coastal plains of Rome and Grosseto between 20,000 and 3,000 years ago. In M. Van Leusen, G. Pizziolo & L. Sarti (eds.), *Hidden Landscapes of Mediterranean Europe*, 161-169. BAR International Series 2320. Oxford: Archeopress.

Attwell, M. & Fletcher, M. 1985. A new technique for investigating spatial relationships: significance testing. In A. Voorrips & S. H. Loving (eds.), *To Pattern the Past: Symposium on Mathematical Methods in Archaeology, Amsterdam 1984*, 181-189. Strasbourg: Conseil de l'Europe.

Attwell, M. R. & Fletcher, M. 1987. An analytical technique for investigating spatial relationships. *Journal of Archaeological Science* 14 (1), 1-11. doi: 10.1016/S0305-4403(87)80002-X

Banning, E. B. 2002. *Archaeological survey*. New York: Plenum Publisher.

Barker, G. 1995a. *A Mediterranean valley: Landscape archaeology and Annales history in the Biferno valley*. London and New York: Leicester University press.

Barker, G. 1995b. Land use and environmental degradation in the Biferno Valley (Central Southern Italy) from Prehistoric times to the present day. In *L'Homme et la dégradation de l'Environnement. XVe Rencontres Internationales de l'Archéologie d'Histoire d'Antibes*, 285 – 297. Juan-les-Pins: APCDA.

Barton, C. M., Ullah, I. I. & Bergin, S. 2010. Land use, water and Mediterranean landscapes: Modelling long-term dynamics of complex socio-ecological systems. *Philosophical Transactions of the Royal Society A*, 368, 5275-5297. doi: 10.1098/rsta.2010.0193

Bintliff, J. L. 1992. Erosion in the Mediterranean lands: A reconsideration of pattern, process and methodology. In J. Boardman & M. Bell (eds.), *Past and present soil erosion*, 125-131. Oxford: Oxbow Books.

Bintliff, J. L. 2000. Landscape change in classical Greece: A review. In F. Vermeulen & M. D. Dapper (eds.), *Geoarchaeology of the landscapes of classical antiquity*, 49-70. Leuven: Peeters Presse.

Bos, R. H. G. & Sevink, J. 1975. Introduction of gradational and pedomorphic features in descriptions of soils. *Journal of Soil Science* 26, 223-233. doi: 10.1111/j.1365-2389.1975.tb01945.x

Brown, A. G. 1997. *Alluvial geoarchaeology: Floodplain archaeology and environmental change*. Cambridge: Cambridge University Press.

Brown, A. G. 1999. Geomorphological techniques in Mediterranean landscape archaeology. In P. Leveau, F. Trément, K. Walsh & G. Barker (eds.), *Environmental reconstruction in Mediterranean landscape archaeology*, 45 – 54. Oxford: Oxbow Books.

Brown, A. G. & Walsh, K. 2017. Societal stability and environmental change: examining the archaeology-soil erosion paradox. *Geoarchaeology: An international journal* 32 (1), 23 – 35. doi: 10.1002/gea.21611

Burri, E., Castiglioni, B. & Sauro, U. 1999. Agriculture, landscape and human impact in some karst areas of Italy. *International Journal of Speleology* 28, 33-54. doi: http://dx.doi.org/10.5038/1827-806X.28.1.3

Butzer, K. W. 2008. *Challenges for a cross-disciplinary geoarchaeology: the intersection between environmental history and geomorphology. Geomorphology* 101, 402 – 411. doi: 10.1016/j.geomorph.2008.07.007

Casarotto, A., De Guio, A., Leonardi, G. & Ferrarese, F. 2011. A GIS-based archaeological predictive model for the study of Protohistoric location-allocation strategies (Eastern Lessinia, VR/VI). *IpoTESI di Preistoria* 4 (2), 1-24. doi: 10.6092/issn.1974-7985/2448

Casarotto, A., Pelgrom, J. & Stek, T. D. 2016. Testing settlement models in the early Roman colonial landscapes of Venusia (291 B.C.), Cosa (273 B.C.) and Aesernia (263 B.C.). *Journal of Field Archaeology* 41 (5), 568-586. doi: 10.1080/00934690.2016.1211474

Cherry, J. F. 1983. Frogs round the pond: Perspectives on current archaeological survey projects in the Mediterranean region. In D. R. Keller & D. W. Rupp (eds.), *Archaeological survey in the Mediterranean area*, 375-416. Oxford: BAR Vol. 155

Citter, C. & Arnoldus-Huyzendveld, A. 2011. *Uso del suolo e sfruttamento delle risorse nella pianura grossetana nel Medioevo: Verso una storia del parcellario e del paesaggio agrario*. Roma: Artemide.

Clarke, D. 1968. *Analytical Archaeology*. London: Methuen.

Coltorti, M. 1983. Le fasi principali dell' evoluzione del paesaggio nel bacino di Isernia (Molise). In M. Coltorti (ed.), *Isernia La Pineta. Un accampamento più antico di 700.000 anni (Isernia, Museo Nazionale)*, 41-47. Bologna: Calderini Editore.

Conolly, J. & Lake, M. 2006. *Geographical information systems in archaeology*. Cambridge: Cambridge University Press.

Cremaschi, M. 1987. *Paleosols and vetusols in the central Po Plain (northern Italy) a study in Quaternary geology and soil development*. Milano: Unicopli Ed.

Cremaschi, M. & Nicosia, C. 2012. Sub-Boreal aggradation along the Apennine margin of the Central Po Plain: geomorphological and geoarchaeological aspects. *Géomorphologie: relief, processus, environnement* 18 (2), 155 – 174. doi: 10.4000/geomorphologie.9810

Cremaschi, M. & Sevink, J. 1987. Micromorphology of paleosol chronosequences on gravelly sediments in Northern and Central Italy. In N. Fedoroff, L.M. Bresson & M.A. Courty (eds.), *Soil Micromorphology*, 577-584. Paris: AFES.

De Guio, A. 1985. Archeologia di superficie ed archeologia superficiale. *Quaderni di Archeologia del Veneto* 1, 176 – 184.

De Guio, A. 2015. Nuove linee di ricerca fra archeologia pre-dittiva e post-dittiva. *Archeologia e Calcolatori* 26, 301-313.

Di Zio, S. & Barnabei, D. 2009. Un modello GIS multicriterio per la costruzione di mappe di plausibilità per la localizzazione di siti archeologici: il caso della costa teramana. *Archeologia e Calcolatori* 20, 309-329.

Diez-Martín, F. 2010. Evaluating the effect of plowing on the archaeological record: the early middle Palaeolithic in the river Duero basin plateaus (north-central Spain). *Quaternary International* 214, 30 – 43. doi: 10.1016/j.quaint.2009.10.024

Eastman, J. R. 2012. *IDRISI Selva manual (Manual Version 17.01)*. Worcester, MA: Clark Labs, Clark University.

Ebert, D. & Singer, M. 2004. GIS, predictive modelling, erosion, site monitoring. *Assemblage* 8.

Esri, 2014. Band Collection Statistics (Spatial Analyst). In *ArcGIS 10.2.2 Help*.

FAO/UNESCO, 1974. *Legend for the Soil Map of the World (1:5,000,000), Volume I - Legend*. Rome: Unesco-Paris.

Feiken, M. 2014. *Dealing with biases: Three Geo-Archaeological approaches to the hidden landscapes of Italy*. Eelde, Netherlands: Barkhuis.

Fentress, E. 2000. What are we counting for? In R. Francovich, H. Patterson & G. Barker (eds.), *Extracting meaning from ploughsoil assemblages*, 45 – 52. Oxford: Oxbow Books.

Finke, P., Harding, J., Sevink J., Gewuster, J. & Stoddart, S. 1994. The dissection of a Bronze and early Iron Age landscape: Site territories and land use. In C. Malone & S. Stoddart (eds.), *Territory, time and state; The archeological development of the Gubbio Basin*, 81 – 93. Cambridge: Cambridge University Press.

Flannery, K. V. 1976. *The early Mesoamerican village*. New York: Academic Press.

Francovich, R., Patterson, H. & Barker, G. (eds) 2000. *Extracting meaning from ploughsoil assemblages*. Oxford: Oxbow Books.

Gillings, M. & Sbonias, K. 1999. Regional survey and GIS: the Boeotia project. In M. Gillings, D. J. Mattingly & J. Van Dalen (eds.), *Geographical information systems and landscape archaeology*, 35-54. Oxford: Oxbow Books.

Given, M. 2004. Mapping and manuring: Can we compare sherd density figures? In S. E. Alcock & J. F. Cherry (eds.), *Side-by-side survey: comparative regional studies in the Mediterranean world*, 13 – 21. Oxford: Oxbow books.

Goodchild, H. 2007. *Modelling Roman agricultural production in the middle Tiber valley, Central Italy*. PhD dissertation, University of Birmingham: eTheses Repository. http://etheses.bham.ac.uk/175/.

Judge, W. J. & Sebastian, L. (eds.) 1988. *Quantifying the present and predicting the past: Theory, method, and application of archaeological predictive modeling*. Denver: US Bureau of Land Management.

Judson, S. 1963. Erosion and deposition of Italian stream valleys during historic time. *Science* 140, 898-899. doi: 10.1126/science.140.3569.898

Kamermans, H. 2000. Land evaluation as predictive modelling: a deductive approach. In G. Lock (ed.), *Beyond the map: archaeology and spatial technologies*, 124-146. Amsterdam: IOS Press.

Kamermans, H. & Sevink, J. 2009. Patterns of middle and upper Palaeolithic land use in Central Lazio (Italy). *Analecta Praehistorica Leidensia* 41, 41-55.

Koopman, A., Kluiving, S., Holdaway, S. & Wendrich, W. 2016. The effects of Holocene landscape changes on the formation of the archaeological record in the Fayum Basin, Egypt. *Geoarchaeology: An International Journal* 31 (1), 17 - 33. doi: 10.1002/gea.21538

Koopmans, T. T. 1980. *Verslag van een bodemkundige kartering, uitgevoerd in het gebied Isernia, Centraal Zuid-Italie*. Master's thesis, University of Amsterdam.

Kvamme, K. L. 1990. One-sample tests in regional archaeological analysis: new possibilities through computer technology. *American Antiquity* 55 (2), 367-381. doi:10.2307/281655

Lang, A. & Bork, H. R. 2006. Past soil erosion in Europe. In J. Boardman & J. Poesen (eds.), *Soil erosion in Europe*, 465-478. Chichester, West Sussex, England: John Wiley & Sons, Ltd. doi: 10.1002/0470859202.ch35

Langston, L. M. 2013. *Site location modeling and prehistoric rock shelter selection on the upper Cumberland plateau of Tennessee.* East Tennessee State University, Electronic Theses and Dissertations. http://dc.etsu.edu/etd/1157

Lefèvre, D., Raynal, J. P., Vernet, G., Kieffer, G. & Piperno, M. 2010. Tephro-stratigraphy and the age of ancient Southern Italian Acheulean settlements: The sites of Loreto and Notarchirico (Venosa, Basilicata, Italy). *Quaternary International* 223 - 224, 360 - 368. doi: 10.1016/j.quaint.2010.02.020

Leonardi, G. 1992a. Assunzione e analisi dei dati territoriali in funzione della valutazione della diacronia e delle modalità del popolamento. In M. Bernardi (ed.), *Archeologia del Paesaggio*, 25 - 66. Firenze: All'Insegna del Giglio.

Leonardi, G. 1992b. *Formation processes and excavation methods in Archaeology: Perspectives*. Padova: Imprimitur.

Nance, J. D. 1983. Regional sampling in archaeological survey: The statistical perspective. *Advances in Archaeological Method and Theory* 6, 289-356.

Potter, T. W. 1976. Valleys and settlements: Some new evidence. *World Archaeology* 8(2), 207-219.

Remmelzwaal, A. 1978. Soil genesis and quaternary landscape development in the Tyrrhenian coastal area of South-Central Italy.

Publicaties van het Fysisch-Geografisch en Bodemkundig Laboratorium van de Universiteit van Amsterdam, no. 28.

Sauer, D. 2010. Approaches to quantify progressive soil development with time in Mediterranean climate-I. Use of field criteria. *Journal of Plant Nutrition and Soil Science* 173 (6), 822–842. doi: 10.1002/jpln.201000136

Sauer, D., Finke, P., Sørensen, R., Sperstad, R., Schülli-Maurer, I., Høeg, H. & Stahr, K. 2012. Testing a soil development model against southern Norway soil chronosequences. *Quaternary International* 265, 18 – 31. doi: 10.1016/j.quaint.2011.12.018

Sbonias, K. 1999. Introduction to issues in demography and survey. In J. L. Bintliff & K. Sbonias (eds.), *Reconstructing past population trends in Mediterranean Europe (3000 B.C. –A.D.1800)*, 1-21. Oxford: Oxbow Books.

Schiffer, M. B. 1987. *Formation Processes of the Archaeological Record.* Albuquerque: University of New Mexico Press.

Sevink, J. 1985. Physiographic soil surveys and archaeology. In C. Malone & S. Stoddart (eds.), *Papers in Italian archaeology IV. The Cambridge conference. Part 1: the Human Landscape*, 41-52. Oxford: BAR International Series 243.

Sevink, J., Vos, P., Westerhoff, W. E., Stierman, A. & Kamermans, H. 1982. A sequence of marine terraces near Latina (Agro Pontino, Central Italy). *Catena* 9 (1 - 2), 361–378. doi: 10.1016/S0341-8162(82)80023-4

Shennan, S. 1988. *Quantifying Archaeology*. Edinburgh: Edinburgh University Press.

Shennan, S., Gardiner, J. & Oake, M. 1985. *Experiments in the Collection and Analysis of Archaeological Survey Data: The East Hampshire Survey*. Sheffield: University of Sheffield.

Siegel, S. 1956. *Nonparametric statistics for the behavioral sciences*. New York: McGraw-Hill.

Stek, T. D., Modrall, E. B., Kalkers, R. A. A., van Otterloo, R. H. & Sevink, J. 2015. An early Roman colonial landscape in the Apennine mountains: Landscape archaeological research in the territory of Aesernia (Central-Southern Italy). *Analysis Archaeologica: An International Journal of Western Mediterranean Archaeology* 1, 229–291.

Tarquini, S., Isola, I., Favalli, M., Mazzarini, F., Bisson, M., Pareschi, M. T. & Boschi, E. 2007. TINITALY/01: a new Triangular Irregular Network of Italy. *Annals of Geophysics* 50 (3), 407 - 425. doi: 10.4401/ag-4424

Tarquini, S. & Nannipieri, L. 2017. The 10 m-resolution TINITALY DEM as a trans-disciplinary basis for the analysis of the Italian territory: Current trends and new perspectives. *Geomorphology* 281, 108-115. doi: 10.1016/j.geomorph.2016.12.022

Tarquini, S., Vinci, S., Favalli, M., Doumaz, F., Fornaciai, A. & Nannipieri, L. 2012. Release of a 10-m-resolution DEM for the Italian territory: Comparison with global-coverage DEMs and anaglyph-mode exploration via the web. *Computers & Geosciences* 38 (1), 168-170. doi: 10.1016/j.cageo.2011.04.018

Terrenato, N. 2000. The visibility of sites and the interpretation of field survey results: towards an analysis of incomplete distributions. In R. Francovich, H. Patterson & G. Barker (eds.), *Extracting meaning from ploughsoil assemblages*, 60-71. Oxford: Oxbow Books.

Terrenato, N. 2004. Sample size matters! The paradox of global trends and local surveys. In S. E. Alcock & J. F. Cherry (eds.), *Side-by-side survey: Comparative regional studies in the Mediterranean world*, 36 – 48. Oxford: Oxbow books.

Terrenato, N. & Ammerman, A. J. 1996. Visibility and site recovery in the Cecina valley survey, Italy. *Journal of Field Archaeology* 23 (1), 91-109. doi: 10.1179/009346996791973990

Torri, D., Borselli, L.,Guzzetti, F., Costanza Calzolari, M., Bazzoffi, P., Ungaro, F., Bartolini, D. & Pilar Salvador Sanchis, M. 2006. Italy. In J. Boardman & J. Poesen (eds.), *Soil erosion in Europe*, 245-261. Chichester, West Sussex, England: John Willey & Sons, Ltd. doi: 10.1002/0470859202.ch20

Toynbee, A. J. 1965. *Hannibal's legacy. 1, Rome and her neighbours before Hannibal's entry*. London: Oxford University Press.

Van Joolen, E. 2003. *Archaeological land evaluation. A reconstruction of the suitability of ancient landscapes for various land uses in Italy focused on the first millennium B.C.*. PhD dissertation, University of Groningen: University of Groningen Research Database. http://hdl.handle.net/11370/78b6a5bf-917d-4a15-969d-230c3c72b690

Van Leusen, M. 1996. Unbiasing the archaeological record. *Archeologia e Calcolatori* 7, 129-135.

Van Leusen, M. 2001. Marginal Landscapes: Survey and interpretation biases in low finds density regions in Italy. In T. Darvill & M. Gojda (eds.), *One Land, Many Landscapes*, 71-74. BAR International Series S987. Oxford: Archaeopress.

Van Leusen, P. M. 2002. *Pattern to process: Methodological investigations into the formation and interpretation of spatial patterns in archaeological landscapes*. PhD dissertation, University of Groningen: University of Groningen Research Database. http://hdl.handle.net/11370/fc9ead23-e2c7-40c5-942f-e06f52198222

Van Leusen, P. M. & Kamermans, H. (eds.) 2005. *Predictive Modelling for Archaeological Heritage Management: a research agenda*. Nederlandse Archeologische Rapporten, vol. 29. Amersfoort: Rijksdienst voor het Oudheidkundig Bodemonderzoek.

Van Leusen, P. M., Pizziolo, G. & Sarti, L. (eds.) 2011. *Hidden Landscapes of Mediterranean Europe. Cultural and methodological biases in pre- and protohistoric landscape studies*. BAR International Series 2320. Oxford: Archaeopress.

Van Otterloo, R. H. 1981. *Soils and Landscape of the Upper-Volturno basin*. Master's thesis, University of Amsterdam.

Van Otterloo, R. H. & Sevink, J. 1983. The Quaternary evolution of the Upper-Volturno basin. In M. Coltorti (ed.), *Isernia La Pineta. Un accampamento più antico di 700.000 anni (Isernia, Museo Nazionale)*, 35-39. Bologna: Calderini Editore.

Van Otterloo, R. H. & J. Sevink 2016. The soil map of the area around Isernia. In T. D. Stek (ed.), *The research activities of the LERC project in Molise* (Internal Report 2016), 21-54. Leiden: Faculty of Archaeology, Leiden University.

Vaughn, S. & Crawford, T. 2009. A predictive model of archaeological potential: an example from northwestern Belize. *Applied Geography* 29 (4), 542-555. doi: 10.1016/j.apgeog.2009.01.001

Verhagen, P. 2007. *Case Studies in Archaeological Predictive Modelling* (ASLU 14). Leiden: Leiden University Press.

Verhoeven, A. A. A. 1991. Visibility factors affecting artifact recovery in the Agro Pontino survey. In A. Voorrips, S. H. Loving & H. Kamermans (eds.), *The Agro Pontino Survey Project: Methods and preliminary results*, 87-97. Studies in Prae- en Protohistorie 6. Amsterdam: University of Amsterdam.

Vermeulen, F. & De Dapper, M. (eds.) 2000. *Geoarchaeology of the landscapes of classical antiquity.* Babesch Supplementa 5. Leuven, Belgium: Peeters.

Vita-Finzi, C. 1969. *The Mediterranean valleys. Geological changes in historical times*. New York: Cambridge University Press.

Vita-Finzi, C. & Judson, S. 1964. Recent deposition and erosion in valleys of the Mediterranean basin. *Geological Society of America Special Paper* 76, 172-173.

Walsh, K. 2014. *The archaeology of Mediterranean landscapes: human-environment interaction from the Neolithic to the Roman period.* New York: Cambridge University Press.

Wheatley, D. & Gillings, M. 2002. *Spatial technology and archaeology. The archaeological applications of GIS*. London & New York: Taylor and Francis.

Zwertvaegher, A. 2012. *The use of integrated process models in a geoarchaeological context. A proof of concept*. PhD dissertation, Ghent University: Ghent University Academic Bibliography. http://hdl.handle.net/1854/LU-3079926

Chapter 5. Surface visibility and legacy survey data: between desktop-based analysis and new fieldwork[16]

Abstract

In archaeological fieldwalking surveys, ground visibility conditions of the walked fields can facilitate or prevent the detection of archaeological sites (*e.g.* Ammerman 1981; Cherry *et al.* 1991; Terrenato & Ammerman 1996; Banning *et al.* 2011; 2017). It is increasingly becoming customary practice in landscape archaeology projects to test survey data for visibility distortions when inferences about the underlying ancient settlement behavior are to be made. This paper proposes such a test for the territory around the ancient town of Venusia (southern Italy), for which the regional *Forma Italiae* survey project produced a large dataset of archaeological sites from 1989 to 2000. We first analyzed whether there is a relationship between surface visibility and the density of sites for the Hellenistic period (*c.* 350 – 50 B.C.) by means of a statistical analysis. Subsequently, we tested the reliability of the site patterns in the legacy data by comparing them with new data recovered from the field in recent re-surveys of this territory, which we carried out from 2013 to 2016 as part of the 'Landscapes of Early Roman Colonization' (LERC) project. We thus assessed the compatibility of the pattern of sites detected by the *Forma Italiae* and the new LERC field surveys. It is especially in this second part of the analysis that we, against more pessimistic estimations, argue that there is no strong correlation between ground survey visibility and the overall regional pattern of discovered legacy sites. We conclude that, on the regional level, legacy survey data are representative and offer significant evidence for the study of ancient settlement patterns.

5.1 INTRODUCTION

The site pattern recorded during regional field survey is not exclusively determined by ancient settlement strategies: there is a multitude of other factors that influence the configuration of the surface record as we see it now in the field (for a synthesis see *e.g.* Cambi & Terrenato 1994, 151-158). In this paper we focus on one important biasing factor: surface visibility as a function of land use and land cover conditions, and we assess whether it affected the number and the pattern of Hellenistic-period archaeological sites registered by the *Forma Italiae* (FI onwards) field survey project in the area around Venosa in southern Italy[17] (Azzena & Tascio 1996; Marchi & Sabbatini

16 This Chapter corresponds to a forthcoming article titled "Surface visibility and legacy survey data: between desktop-based analysis and new fieldwork" by Anita Casarotto, Jesús García Sánchez, Tesse D. Stek & Jeremia Pelgrom , submitted to *FOLD&R: the Journal of Fasti Online* (http://www.fastionline.org/folder.php?view=home).
The LERC fieldwork in the territory of Venosa and surrounding municipalities (campaigns 2013 - 2017) was financially supported by NWO (Netherlands Organization for Scientific Research), Leiden University, and the KNIR (Royal Netherlands Institute in Rome), and was coordinated by Dr. Jeremia Pelgrom (Field project Director), in collaboration with Dr. Tesse D. Stek, Dr. Jesús García Sánchez (campaigns 2014 - 2017), Anita Casarotto (campaigns 2013 - 2015), and Prof. Maria Luisa Marchi. We would like to thank all field walkers and pottery specialists who took part in the various LERC survey campaigns in Venosa: Lucia Lecce (pottery specialist), Arthur Hamel and Dr. Giancluca Cantoro (experts in geophysics, drone devices, and aerial photogrammetry), Miriam Alba Luzón, Frank Beijaard, Jochem de Boer, Beatrijs de Groot, Mike de Heij, Rani Evaert, David Fernández Abella, Annachiara Fiore, Kristel Henquet, Stefan Kooi, Iris Kramer, Bruno Laso Gálvez, Bert Melkebeek, Serban Micu, Koos Mol, Ben Naylor, Zizi Rico Neves, Bianca Olteanu, Ine de Pape, Isabel Perales Pérez, Manuel Peters, Jorge Santamaría Treceño, Jacqueline Splinter, Eef Stoffels, Niels Stoffels, Nele Struyf, Ivar Svensson, Sofia Taipale, Leon Theelen, Eduardo Arancón Torrecilla, Vanesa Trevín Pita, Ties Verhoeven, Eva Visser, Naomi Vlieks, and Evelien Witmer. We are also grateful to Soprintendenza Archeologia Belle Arti e Paesaggio della Basilicata, the staff of Museo Archeologico Nazionale di Venosa, and to Prof. Maria Luisa Marchi who kindly gave us access to the *Forma Italiae* data. Finally, we sincerely thank the community of Venosa for their warm hospitality.

17 For methodological information on this survey, reference is made to section 2.2 of Chapter 2 and 3.2 of Chapter 3.

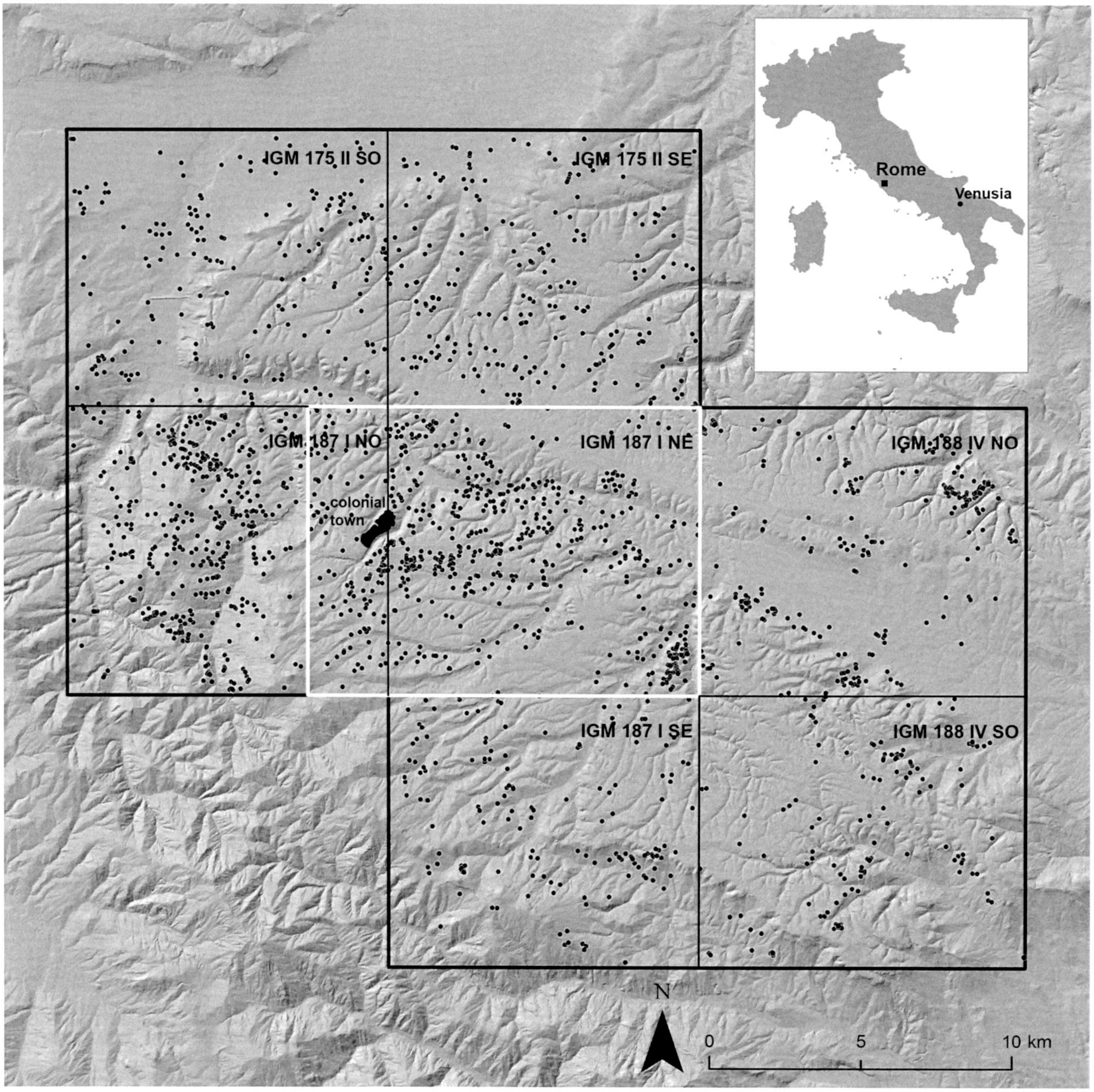

Fig. 5.1 *Forma Italiae* site dataset for the territory surrounding the ancient town of Venusia. In total, 1899 archaeological sites (dating from Prehistory to the Middle Ages) were recorded in a territory of ca. 700 sq km (Marchi, Sabbatini 1996; Sabbatini 2001; Marchi 2010). The labels indicate the IGM maps (Istituto Geografico Militare, 1:25,000) used to register archaeological sites. The raster base map is the shaded relief calculated from the 10 m-resolution DEM named TINITALY/01 (Tarquini *et al.* 2007; 2012; Tarquini& Nannipieri 2017).

1996; Sabbatini 2001; Marchi 2010; Marchi 2016) (Fig. 5.1).

This paper consists of two parts. In the first part the site density recorded by the FI survey project is tested. Possible correlations between survey visibility conditions and the number of Hellenistic sites were tested through a linear regression analysis. The aim of this approach was to evaluate whether localized high densities of sites (registered by the original FI survey on distribution maps as point clusters) significantly occur in zones with good visibility conditions and if, instead, low site density or empty spaces significantly appear in zones with low visibility conditions.

In the second part, we designed a special testing strategy to analyze the pattern of the Hellenistic settlements (see also García Sánchez *et al.* 2017). In previous investigations (Chapters 2 and 3), it was demonstrated that the spatial pattern behind the Hellenistic settlement distribution is clustered. Here, the aim is to test whether this pattern is reliable or, instead, affected by survey visibility conditions. To test the validity of the settlement distribution recorded some decades ago by the FI team, we carried out re-surveys in the Venosa area, as part of the LERC project (Landscapes of Early Roman Colonization, campaigns 2013 – 2016)[18] (for a similar approach see Seubers & Tol 2016; see also Ammerman *et al.* 2013). Our re-surveys targeted those landscape areas with strong clustering of Hellenistic sites as well as low density and entirely empty zones. "Ground truthing" the original dataset in these distinctive zones, allowed us to establish whether the site patterns and processes documented in the FI survey could still be seen, or if they instead had changed.

5.2 DESKTOP-BASED ANALYSIS: TESTING THE LEGACY SITE DENSITY

In this section we ask whether the density of Hellenistic settlements has been affected by surface visibility conditions. For this analysis, we use the visibility map of the territory close to the ancient town of Venusia, which is based on land use and land cover factors (*c.* 120 sq. km, contoured with a white rectangle in Fig. 5.1) (Azzena & Tascio 1996; Marchi & Sabbatini 1996, 107). This map proved to be powerful for the statistical analysis presented in this paper (for another useful application see Chapter 2). Within said sample area, *c.* 600 archaeological sites were recorded by the FI team, of which 262 settlements were generally dated to the Hellenistic period (350 -50 B.C.) (Marchi & Sabbatini 1996) (Fig. 5.2).

Using a linear regression analysis, we assessed whether there was a significant correlation between site agglomerations and survey units in good visibility conditions and, conversely, if less sites were attested in units with less optimal visibility conditions (for the same approach see Terrenato & Ammerman 1996).

Since no information is available about the extension and/or the shape of the original survey units (Pelgrom *et al.* 2014, 33 – 34), we created an arbitrary sampling grid composed by units of 1 sq km (Fig. 5.3). We then tested for each and every unit whether a direct proportional relationship existed between the extension of the surveyed surface in good visibility conditions (*i.e.* class 5)[19] and the number of Hellenistic sites recorded. If a significant relationship between the two factors exists, one expects that the more surface is well visible within each unit, the higher the number of sites recorded in these units should be.

As we show in Fig. 5.4 however, we did not find a direct proportional relationship between the percentage of area with good visibility and the number of Hellenistic-period sites documented by

18 Funds were provided by NWO (Netherlands Organization for Scientific Research) and the KNIR (Royal Netherlands Institute in Rome). See also Stek & Pelgrom 2013.

19 It is important to note that the tiny and localized zone in 'class 6' (optimal visibility, 0.48% of the territory) (see Fig. 5.2, p. 94) corresponds to a small land plot where a vineyard was in the process of being planted precisely when the survey took place there in July 1988 (Marchi & Sabbatini 1996, 113, footnote 134). Sites in this zone were recorded under exceptional conditions (while sites were being dug up from the subsoil and destroyed immediately after when the field was leveled and prepared for the vineyard plantation). This site sample is thus unrepresentative for the process of recording sites at the surface through field walking. For this reason, in the following analyses, class 6 is merged with class 5, corresponding to the 'good visibility' land type.

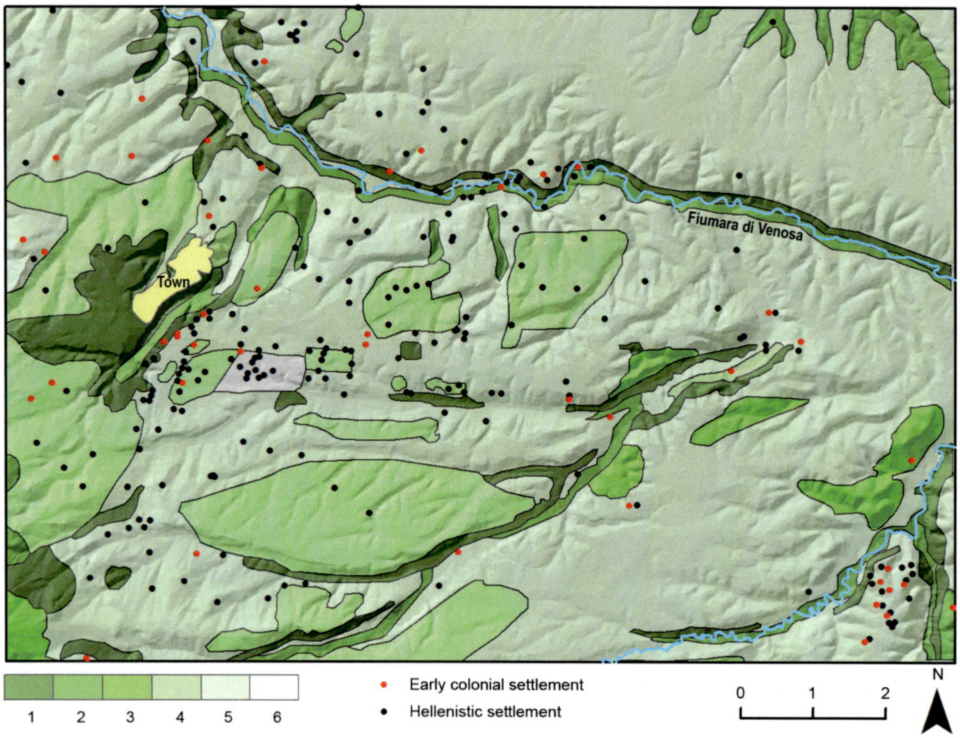
Town
Fiumara di Venosa
1 2 3 4 5 6
Early colonial settlement
Hellenistic settlement
0 1 2
N

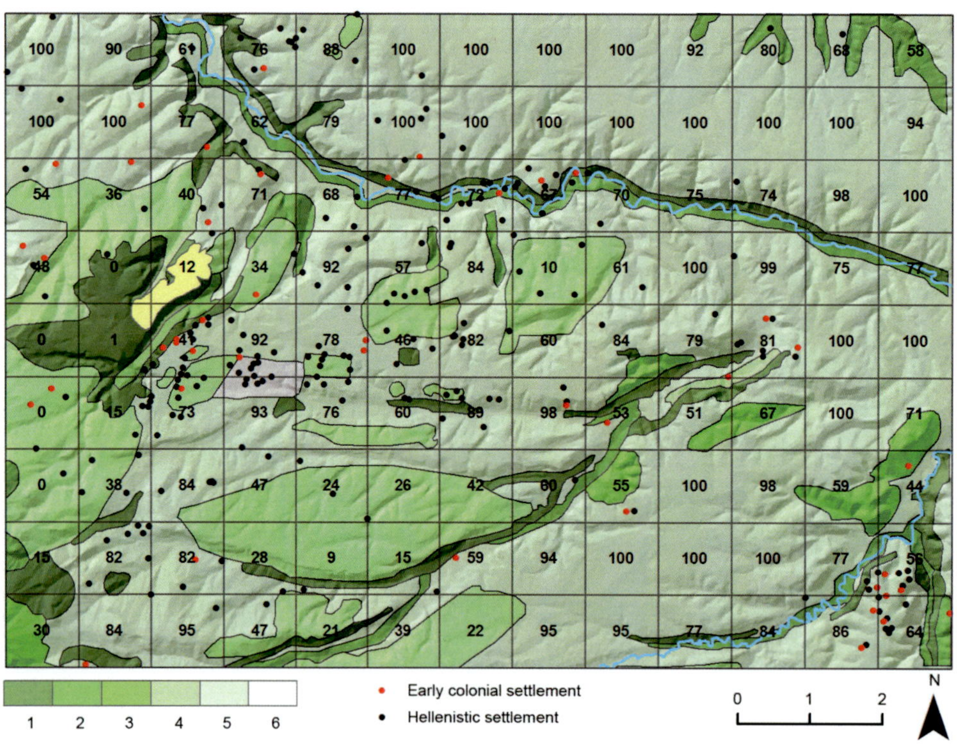
100 90 61 76 88 100 100 100 100 92 80 68 58
100 100 77 62 79 100 100 100 100 100 100 100 94
54 36 40 71 68 77 72 67 70 75 74 98 100
48 0 12 34 92 57 84 10 61 100 99 75 77
0 1 41 92 78 46 82 60 84 79 81 100 100
0 15 73 93 76 60 89 98 53 51 67 100 71
0 38 84 47 24 26 42 60 55 100 98 59 44
15 82 82 28 9 15 59 94 100 100 100 77 56
30 84 95 47 21 39 22 95 95 77 84 86 64
1 2 3 4 5 6
Early colonial settlement
Hellenistic settlement
0 1 2
N

◀ **Fig. 5.2** Hellenistic settlement site distribution and surface visibility map (scale 1: 200,000, based on Marchi, Sabbatini 1996, 107; Azzena & Tascio 1996. Graphic elaboration by Anita Casarotto) for the area close to the town, which corresponds to the area contoured by a white rectangle in Fig. 5.1. Visibility values range from 1 (very low visibility) to 6 (optimal visibility) (see also Azzena & Tascio 1996). The raster base map is the shaded relief calculated from the 10 m-resolution DEM named TINITALY/01 (Tarquini *et al.* 2007; 2012; Tarquini & Nannipieri 2017).

◣ **Fig. 5.3** Arbitrary grid composed by units of 1 sq km. The numeric labels indicate the percentage of the unit area in 'good visibility' conditions (class 5). The raster base map is the shaded relief calculated from the 10 m-resolution DEM named TINITALY/01 (Tarquini *et al.* 2007; 2012; Tarquini & Nannipieri 2017).

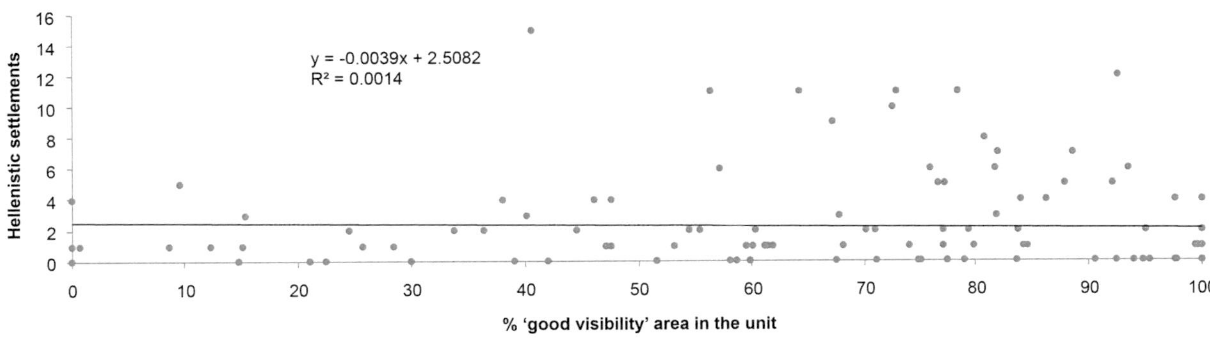

Fig. 5.4 Regression analysis on the Hellenistic settlements. The correlation coefficient (r) is -0.04. The least-squares method is applied to determine the best fit line to data (see also Terrenato & Ammerman 1996).

the FI survey. This result therefore seems to support the accuracy of the number of recorded sites, and also of their pattern in space, at least on this scale of analysis. However, it must be noted that the scale of the visibility map does not provide the level of detail needed to allow a finer analysis of visibility distortions (see discussion in Chapters 2 and 4). Higher resolution visibility maps are not available at the moment; once more detailed maps for assessing smaller-scale visibility distortions would become accessible, the results from this regression analysis may likely change. In light of these limitations, an additional test of a possible correlation between visibility and site patterns was carried out (see below).

5.3 CONTROL SAMPLES FROM NEW FIELDWORK: TESTING THE LEGACY SITE PATTERN

Within the context of the LERC project (Stek & Pelgrom 2013; Stek *et al.* 2016), 3283 ha of the territory of Venosa (PZ, Basilicata) and its surrounding municipalities have been systematically re-surveyed in the falls of 2013 to 2016[20] (Fig. 5.5). As the original FI survey, the LERC survey also worked with teams of five field walkers spaced ten m apart in line transects in each field unit, and set a shard density threshold of five fragments per square meter to identify archaeological sites at the surface.

20 Such re-visits offered the opportunity to assess the state of preservation of the legacy survey record. It was sadly acknowledged that mechanized agricultural activities (vegetation clearance, tillage, and land leveling) and water erosion (Torri *et al.* 2006; Torri & Borselli 2011), are triggering the disappearance of the archaeological record both at the surface and beneath it. In particular, modern plantations of vineyards, orchards, and olive trees, which require land leveling and extensive sediment movement, strongly affect the soil stratigraphy in this region (Borselli *et al.* 2006), and thereby the archaeology contained in it. This dramatic situation has called for the monitoring of the geomorphic modifications caused by unregulated plowing and land leveling activities. EU funds were recently budgeted to monitor and prevent the widespread degradation of this landscape, and to foster sustainable territorial planning strategies for the preservation of the soil (see the DESIRE project and the Rendina case-study: http://desire-project.eu/ , Torri *et al.* 2012).

Fig. 5.5 LERC field survey activities in the territory of the Latin colony of Venusia (LERC survey 2015) (photo by Anita Casarotto).

These sites were mapped in the field by means of GPS positioning systems and mobile devices (*i.e.* tablets), and in the laboratory they were integrated through the GIS platform.

As noted, the LERC team aimed to verify the reliability of the overall large-scale clustered pattern documented by the legacy Hellenistic settlement distribution (cf. Chapter 3) by means of site-oriented field survey. The strategy adopted for this test can be defined as a local "ground-truthing" of *a.* localized clusters of sites, *b.* lower density zones, and *c.* empty areas in-between these clusters. Several of the detected site clusters were classified by the FI team as large nucleated villages. The LERC re-visits confirmed these interpretations and also documented significant new information about the dating and the spatial configuration of these large rural settlements (Pelgrom *et al.* 2014; 2016).

For the aim of this paper, we focus here on three of the control sample areas re-surveyed by the LERC team (shown in Fig. 5.6). The first sample area (A - Messero – Lo Scannato) was chosen because several high and localized site concentrations (clusters) were originally documented by the FI survey here (these point clusters have a Hellenistic site density equal to or greater than five sites per sq km, see Chapter 3). The second sample (B - Salto dei Paladini/Lasano) had, in contrast, a remarkably low density of Hellenistic settlements (an average of one or two Hellenistic sites per sq km). The third sample area was devoid of sites altogether (C - Lì Castellani) and located between the landscape zones with clusters of sites.

In selecting the sample areas, we also took into account the visibility conditions. We selected those areas with comparable visibility conditions during

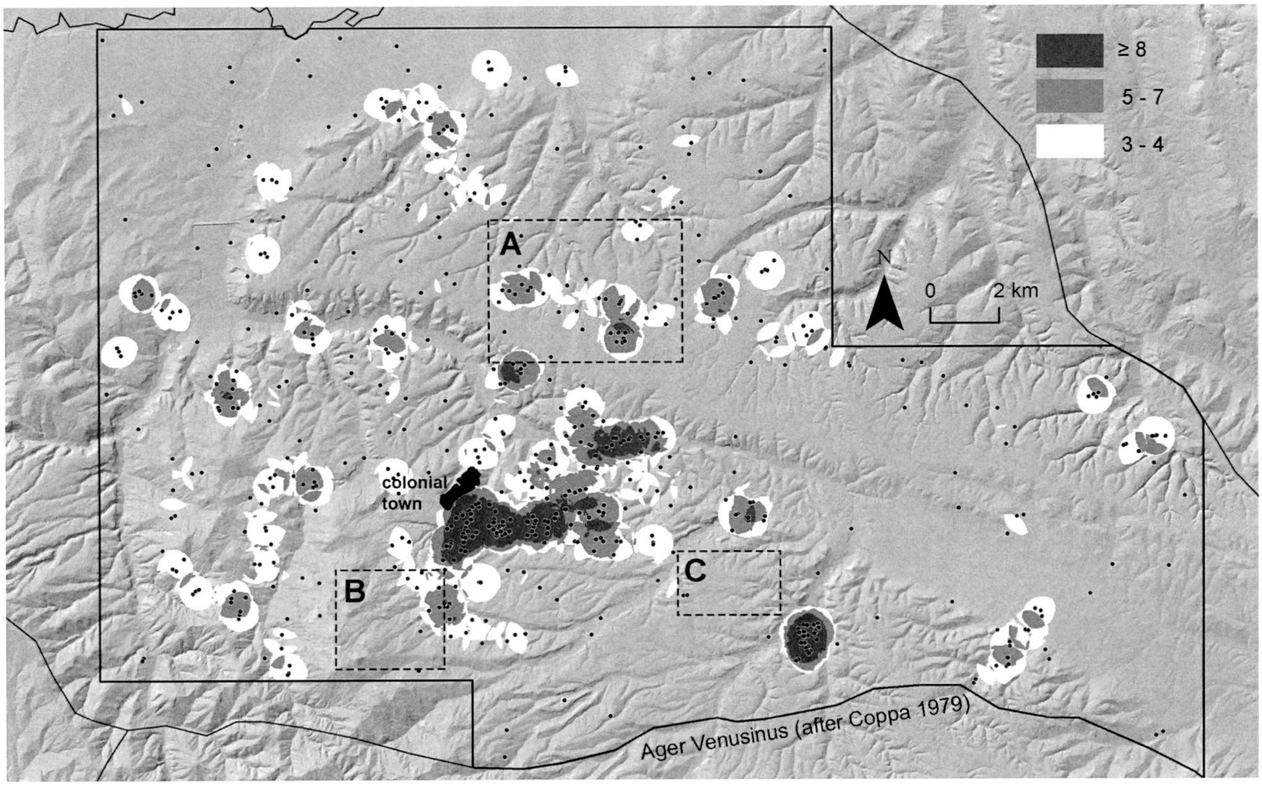

Fig. 5.6 Location of the three control sample areas re-surveyed by the LERC team. The palette indicates the density of legacy Hellenistic settlement sites recorded by the FI team (see also Fig. 3.2 in Chapter 3, p. 50). The raster base map is the shaded relief calculated from the 10 m-resolution DEM named TINITALY/01 (Tarquini *et al.* 2007; 2012; Tarquini & Nannipieri 2017).

the previous FI surveys and our re-surveys (for the FI surveys, we used the published FI visibility maps: Marchi & Sabbatini 1996, 107; Sabbatini 2001, 59). Moreover, the Corine Land Cover 1990, 2000, 2006, 2012 (1: 100,000)[21], the map Carta dell'Uso Suolo della Regione Basilicata 2013 (1: 5000)[22], and the Coltura Agricola 2006 map (1: 5000)[23], indicate that the land use in these areas did not remarkably change over the last decades (*i.e.* mainly non-irrigated arable land, with associated sporadic vineyards, orchards, and other crops). Thanks to this continuity over time in visibility and land use conditions, we could theoretically exclude that major variations in surface conditions (occurring in the last decades, after the FI survey and before the LERC survey) were responsible for possible differences in site patterns recorded by the two surveys at two different moments in time.

Fig. 5.7 clearly shows that the LERC data patterns are comparable to the FI data patterns in all three sample zones, in both empty and densely settled areas. In sample area A (surveyed area 748 ha) (Fig. 5.8), the LERC team recorded localized site densities (clusters) in the same zones where the FI team did several decades before. Many LERC sites were even found in the precise position and extension with the legacy sites. Additionally, some

21 European Environment Agency (EEA). https://www.eea.europa.eu/publications/COR0-landcover

22 Regione Basilicata - Centro cartografico dipartimentale della Direzione Generale del Dipartimento Ambiente e Territorio, Infrastrutture, Opere pubbliche e Trasporti. https://rsdi.regione.basilicata.it

23 Regione Basilicata - Centro cartografico dipartimentale della Direzione Generale del Dipartimento Ambiente e Territorio. https://rsdi.regione.basilicata.it

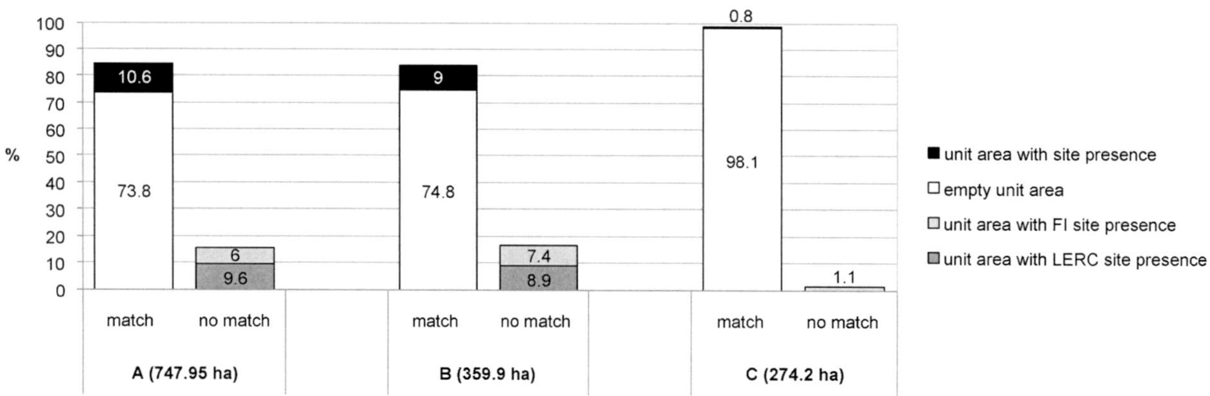

Fig. 5.7 Percentages of matching and non-matching re-surveyed area in each control sample zone (A, B, and C).

Fig. 5.8 Comparison between the LERC site pattern and the FI site pattern in sample area A. The material scatters documented by the LERC team are indicated as well (black polygons).

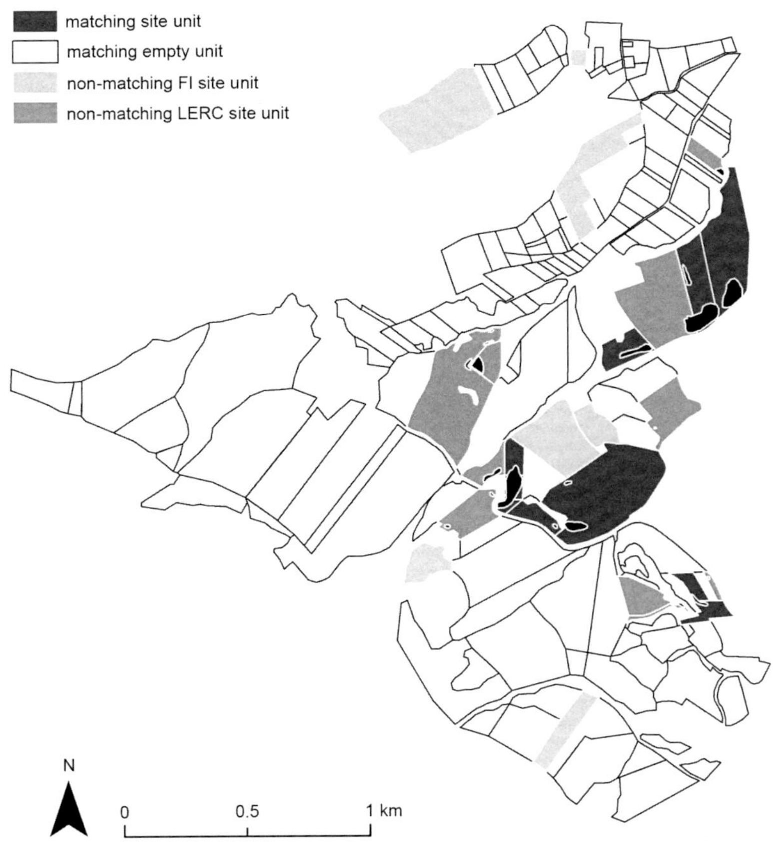

Fig. 5.9 Comparison between the LERC site pattern and the FI site pattern in sample area B. The material scatters documented by the LERC team are indicated as well (black polygons).

new sites were identified[24], and a few others were not found, probably because these sites were destroyed by recent intensive agricultural activities.

The same effect is documented for sample area B (surveyed area 360 ha) (Fig. 5.9). Also in this case, there was a strong correlation between the pattern mapped by the LERC re-surveys and the pattern recorded by the original FI survey.

Particularly interesting is the comparison between the legacy pattern and LERC pattern in sample area C, the 'empty' area (Fig. 5.10). The LERC team adopted an intensive survey strategy for mapping an area of 274 ha. Within rectangular units of *c.* 2500 sq m in size, sherds and other finds were collected by surveyors along each walking line. GPS and tablets were used in this case as well to record the position and the extension of archaeological sites. Although working in different time periods and under perfect visibility conditions (clear, arable fields), both teams found little to nothing: just two scatters of material

24 It is possible that some of these new sites are the same sites recorded by the FI team, but shifted on the map one or few dozens of meters away from their original position. If that is the case, this shift in mapped position can be explained in two ways. First, it is possible that topsoil movements within the field induced by plowing, land leveling, and shallow landslides had moved the original sites (or, alternatively, had unearthed new material, thus creating new sites). Second, the different scale adopted by the two surveys to register sites can explain the "imperfect" match in the mapped position between some FI and LERC sites. In this analysis we used the 1:25,000 scale distribution maps produced by the FI team (Marchi & Sabbatini 1996; Sabbatini 2001; Marchi 2010) as a support to digitalize in GIS the position of the legacy sites. GPS coordinates were used, instead, to map the position of the LERC sites. To incorporate this discrepancy in mapping resolution techniques, field units, rather than the point locations of the FI and LERC sites, were considered when comparing the two datasets (see Figs. 5.8-10).

Fig. 5.10 Comparison between the LERC site pattern and the FI site pattern in sample area C. The material scatters documented by the LERC team are indicated as well (black polygons).

were detected at the edge of this empty zone (by both teams, in the same area, with comparable extension), which were interpreted as the remains of a Republican period settlement.

5.4 CONCLUSIONS

The two types of analysis carried out in this paper indicate that the site patterns recorded by field survey are not just the product of distortions by visibility factors. This underlines the value of legacy survey data for reconstructing ancient settlement organization and historical developments on a regional scale (Chapter 3 and 6). The statistical, desktop-based approach confirms the reliability of the number of sites registered by the FI team, and new fieldwork conducted by the LERC team indicates that there is no significant relationship between the recorded clustered site pattern and surface visibility. There is instead a significant match between the legacy site pattern and the survey data recently collected by the LERC team. Especially this latter, field-based probing shows that the site pattern exhibited by legacy site survey data in this area is not just the result of surface visibility distortions (see discussion in Barker 1991).

Of course, due to the generally problematic nature of survey data (Lloyd & Barker 1981; Ammerman 1985; Pasquinucci & Trement 1999; Francovich *et al.* 2000; Fentress 2000), and the difference in resolution in mapping strategies adopted by the two surveys (cf. footnote 24, p. 99), we cannot expect that the sites recorded by the FI team are the very same sites recorded by the LERC team; this is not our point here. It is significant, however, that on a regional scale the macro patterns produced by the FI and LERC surveys are very similar, and lead to similar historical reconstructions of ancient settlement in this area.

It is seminal to underscore, however, that, in line with our research questions, for this analysis we only selected areas with good visibility and minimal impact by modern anthropogenic practices following the FI survey, with thus optimal preservation of the archaeological record. The results will of course be very different for those areas that, after the FI survey, have been subjected to strong human interventions. In the Venosa area, there are zones that have been drastically modified during the recent years of intensive agricultural activity. An example is the land where extensive permanent crops such as vineyards and olive orchards were planted recently (Borselli *et al.* 2006)[25]. While surveying those portions of the landscape, it became clear that the surface record registered only a few decades ago is disappearing rapidly. There are large zones where modern agricultural practices are unsustainable for the preservation of the soil and the archaeological record on and under the surface. Erosion and desertification are tangible threats in these areas. Despite the existence of still-preserved patches where archaeological information can as yet be retrieved, the entire archaeological landscape of the Venosa area is currently under threat. If no action is taken to control unregulated farming practices it is only a matter of time before the whole archaeological landscape will be lost forever (see discussion in section 1.2 of Chapter 1).

The FI dataset represents a unique source of information for the investigation of ancient settlement patterns. The FI team was likely able to record sites at the surface immediately before the rise of both a massive production of wine and olive oil and the destructive activities related to intensive mechanized agriculture. This legacy dataset is the most representative evidence that can ever be recorded for this landscape, and thus constitutes an invaluable archive for studying the past.

25 Detailed and extensive analyses aimed at studying the relationship between modern agrarian practices and the recording of the LERC data in Venosa are planned for the future.

5.5 REFERENCES

Ammerman, A.J. 1981. Surveys and archaeological research. *Annual Review of Anthropology* 10, 63-88.

Ammerman, A.J., 1985. Plow-zone experiments in Calabria, Italy. *Journal of Field Archaeology* 12(1), 33-40.

Ammerman, A.J., Koster, H., Pfenning, E. 2013. The longitudinal study of land-use at Acconia: Placing the fieldwork of the survey archaeologist in time. *Journal of Field Archaeology* 38(4), 291-307. doi: 10.1179/0093469013Z.00000000060

Azzena G., & Tascio, M. 1996. Il Sistema Informativo Territoriale per la carta archeologica d'Italia. In M.L. Marchi & G. Sabbatini, *Venusia (IGM 187 I NO / I NE)* (Forma Italiae 37), 281 – 297. Firenze: Leo S. Olschki.

Banning, E.B., Hawkins, A. L. & Stewart, S. T. 2011. Sweep widths and the detection of artifacts in archaeological survey. *Journal of Archaeological Science* 38 (12), 3447-3458. doi: 10.1016/j.jas.2011.08.007

Banning, E.B., Hawkins, A.L., Stewart, S.T., Hitchings, P. & Edwards, S. 2017. Quality Assurance in Archaeological Survey. *Journal of Archaeological Method and Theory* 24(2), 466 – 488. doi: 10.1007/s10816-016-9274-2

Barker, G. 1991. Approaches to archaeological survey. In: G. Barker & J. Lloyd (eds.), *Roman landscapes: Archaeological survey in the Mediterranean region*, 1 – 9. London: British School at Rome.

Borselli, L., Torri, D., Øygarden, L., De Alba, S., Martinez-Casasnovas, J.A., Bazzoffi, P. & Jakab, G. 2006. Land Levelling. In: J. Boardman & J. Poesen (eds.), *Soil Erosion in Europe*, 643-658. West Sussex, England: John Willey & Sons Ltd.

Cambi, F. & Terrenato, N. 1994. *Introduzione all'archeologia dei paesaggi*. Roma: La Nuova Italia Scientifica.

Cherry, J. F., Davis, J. L. & Mantzourani, E. 1991. *Landscape archaeology as long-term history. Northern Keos in the Cycladic Islands from Earliest settlement until modern times*. Los Angeles: UCLA Institute of Archaeology.

Fentress, E. 2000. What are we counting for? In R. Francovich, H. Patterson & G. Barker (eds.), *Extracting meaning from ploughsoil assemblages*, 45 – 52. Oxford: Oxbow Books.

Francovich, R., Patterson, H. & Barker, G. (eds.) 2000. *Extracting meaning from ploughsoil assemblages*. Oxford: Oxbow Books.

García Sánchez, J., Pelgrom, J. & Stek, T.D. 2017. Comparing re-surveys in Isernia and Venosa (Molise and Basilicata, Italy). *Mediterranean Archaeology and Archaeometry: International Scientific Journal* 17(3), 39-52. doi: 10.5281/zenodo.1005454

Lloyd, J. A. & Barker, G. 1981. Rural settlement in Roman Molise: problems of archaeological surveys. In G. Barker & R. Hodges (eds.), *Archaeology and Italian Society*, British Archaeological Reports, International Series 102, Oxford, 289 - 304.

Marchi, M.L. 2010. *Ager Venusinus II (IGM 175 II SO; 187 I NO; 187 I SE; 188 IV NO; 188 IV SO)* (Forma Italiae 43). Firenze: Leo S. Olschki.

Marchi, M.L. 2016. Carta Archeologica d'Italia – "Forma Italiae" project: Research method. In *LAC 2014 Proceedings (Rome 2014)*. doi: http://dx.doi.org/10.5463/lac.2014.42

Marchi, M.L. & Sabbatini, G. 1996. *Venusia (IGM 187 I NO / I NE)* (Forma Italiae 37). Firenze: Leo S. Olschki.

Pasquinucci, M., Trément, F. 1999 (eds.), *Non-destructive techniques applied to landscape archaeology*. The archaeology of Mediterranean Landscapes 4. Oxford.

Pelgrom, J., Lecce, L., Garcia Sanchez, J. & Stek, T.D. 2016. Le indagini olandesi sul pianoro di Masseria Casalini Sottana (2013-2014). In A. De Siena & T. Giammatteo (eds.), *Palazzo San Gervasio: Modalità insediative e pratiche funerarie dal territorio*, 277-288. Venosa: Osanna Ed.

Pelgrom, J., Marchi, M.L., Cantoro, G., Casarotto, A., Hamel, A., Lecce, L., García Sánchez, J. & Stek, T.D. 2014. New approaches to the study of village sites in the territory of Venosa in the Classical and Hellenistic periods. *Agri Centuriati* 11, 31-60. doi: 10.1400/233577

Sabbatini, G. 2001. *Ager Venusinus I. Mezzana del Cantore (IGM 175 II SE)* (Forma Italiae 40). Firenze: Leo S. Olschki.

Seubers, J.F. & Tol, G.W. 2016. City, country and crisis in the Ager Crustuminus. Confronting legacy data with resurvey results in the territory of ancient Crustumerium. *Palaeohistoria* 57/58 (2015-2016), 137-155.

Stek, T.D., Pelgrom, J., Casarotto, A., García Sánchez, J., Götz, L., Hamel, A., Iannantuono, K., Kalkers, R.A.A., Termeer, M.K. & Waagen, J. 2016. Santuari, villaggi, centri fortificati e prima urbanizzazione tra Sanniti e Romani: Nuove ricerche archeologiche nell'Alta Valle del Tappino, nel territorio della colonia latina di Aesernia e in quello della città frentana di Larinum, *ArcheoMolise* 26, 24-39.

Stek, T. D. & Pelgrom, J. 2013. Landscapes of Early Roman Colonization: Non-urban settlement organization and Roman expansion in the Roman Republic (4th–1st centuries B.C.). *Tijdschrift voor Mediterrane Archeologie* 50, 87.

Tarquini, S. & Nannipieri, L. 2017. The 10 m-resolution TINITALY DEM as a trans-disciplinary basis for the analysis of the Italian territory: Current trends and new perspectives. *Geomorphology* 281, 108-115. doi: 10.1016/j.geomorph.2016.12.022

Tarquini, S., Isola, I., Favalli, M., Mazzarini, F., Bisson, M., Pareschi, M. T. & Boschi, E. 2007. TINITALY/01: a new Triangular Irregular Network of Italy. *Annals of Geophysics* 50 (3), 407 - 425. doi: 10.4401/ag-4424

Tarquini, S., Vinci, S., Favalli, M., Doumaz, F., Fornaciai, A. & Nannipieri, L. 2012. Release of a 10-m-resolution DEM for the Italian territory: Comparison with global-coverage DEMs and anaglyph-mode exploration via the web. *Computers & Geosciences* 38 (1), 168-170. doi: 10.1016/j.cageo.2011.04.018

Terrenato, N. & Ammerman, A. J. 1996. Visibility

and site recovery in the Cecina valley survey, Italy. *Journal of Field Archaeology* 23 (1), 91-109. doi: 10.1179/009346996791973990

Torri, D. & Borselli, L. 2011. Water Erosion. In Pan Ming Huang, Yuncong Li & Malcolm E. Sumner (eds.), *Handbook of soil sciences: Resource management and environmental impacts*, ch.22, 1-19. Boca Raton FL: CRC Press Taylor & Francis

Torri, D., Borselli, L.,Guzzetti, F., Costanza Calzolari, M., Bazzoffi, P., Ungaro, F., Bartolini, D. & Pilar Salvador Sanchis, M. 2006. Italy. In J. Boardman & J. Poesen (eds.), *Soil erosion in Europe*, 245-261. Chichester, West Sussex, England: John Willey & Sons, Ltd.

Torri, D., Borselli, L. & Lollino, P. 2012. Rendina study site, Basilicata - Italy. Highlights of the work carried out in the DESIRE project. Based on work by University of Basilicata, Italy. http://www.desire-his.eu/en/rendina-italy/840-highlight-conclusions

Chapter 6. A systematic GIS-based analysis of settlement developments in the landscape of Venusia in the Hellenistic-Roman period[26]

Abstract

This paper investigates the settlement developments of the landscape around the ancient town of Venusia in southern Italy using legacy field survey data. A Latin colony was established here in 291 B.C., and also other subsequent Roman colonization movements are known from the literary sources. As in many other Roman colonial landscapes, trends in the settlement data of Venusia have previously been linked to the impact of Roman colonization, which is usually understood as a drastic transformation of the pre-Roman settlement landscape and land use. Rather than using theories on Roman colonial strategies for explaining possible settlement patterns (deductive approach, see Chapter 3), this paper presents an alternative, descriptive, bottom-up approach, and GIS-based inductive location preference analysis to investigate how the settlement landscape evolved in the Hellenistic and Roman periods (particularly in the 4th-1st century B.C.). Following closely the settlement choices from the pre-Roman conquest period onwards and assessing patterns in continuity and change in the settlement record, we demonstrate that pre-Roman rural settlement and land use strategies were not eradicated but instead strongly determined the location preferences for later settlements in the 'colonial' periods. If these settlement trends can be related at all to the colonization waves mentioned in the ancient literary sources, the conclusion should be that Roman colonization did not lead to radical landscape and land use transformations, as has traditionally been suggested. Instead, an organic and complementary rural infill over time is documented, in which cultural factors instead of land use potential played a key role.

6.1 INTRODUCTION

The formation of early Roman colonial landscapes has often been reconstructed using literary evidence on Roman colonization strategies, which for the most part was written after the Civil Wars (1st century B.C.). Several recent studies have stressed the high risk of anachronism underlying such an approach, and have highlighted the dangers of adopting synchronic text-based approaches to understand previous mid-Republican colonization practices (*e.g.* Crawford 1995; Torelli 1999; Bispham 2006; Patterson 2006; Pelgrom & Stek 2014). In this paper a different approach to understanding colonial settlement strategies is offered, using a GIS-based quantitative and qualitative analysis of settlement behavior and location preferences in the colonial landscape of Venusia. This paper complements and further expands the research strategy outlined in a set of previous articles, which have, instead, focused on settlement pattern analysis and deductive reasoning (Casarotto *et al.* 2016, *i.e.* Chapter 3) and survey methodological issues (Casarotto *et al.* 2018, *i.e.* Chapter 4 and 5). As such, it offers an approach to use legacy, site-based datasets for understanding settlement patterns. The aim here is to move from point observations to area-based interpretations of the ancient settlement processes underpinning site configurations, to get a

26 This Chapter corresponds to the article "A systematic GIS-based analysis of settlement developments in the landscape of Venusia in the Hellenistic-Roman period" by Anita Casarotto, Jeremia Pelgrom & Tesse D. Stek, originally published (25 November 2017) in *Archaeological and Anthropological Sciences* (publisher Springer) as an Open Access article distributed under the terms of the Creative Commons Attribution 4.0 International License (https://creativecommons.org/licenses/by/4.0/). This article is accessible through this link: https://doi.org/10.1007/s12520-017-0561-y
The authors of the article wish to thank the Soprintendenza Archeologia Belle Arti e Paesaggio della Basilicata, as well as Prof. dr. Maria Luisa Marchi (University of Foggia) for their support.

firm grip on the evolution and morphogenesis of the landscape as a whole (see also discussion in De Guio 1985).

The territory around Venusia presents a particularly rich case-study to investigate settlement developments in relation to Roman colonization. Thanks to the work by M.L. Marchi and G. Sabbatini (Marchi & Sabbatini 1996; Sabbatini 2001; Marchi 2010), it is one of the best studied Roman colonial territories in Italy. Additionally, it has, according to the literary sources, witnessed a whole series of colonial settlements. At least in theory, this offers the opportunity to assess possible changes in settlement strategies in relation to Roman colonialism over time. In the 3rd century B.C., as part of its expansionistic enterprise taking place across the entire Italian peninsula, Rome sent out the first wave of colonists to settle in this territory. Subsequently, other groups of colonists are reported to have been sent there, in 200 B.C. (allegedly to repopulate the region after the Second Punic War), at the end of the 2nd century B.C. (as a response to the Gracchan land reforms), and later on in the Triumviral period (perhaps Augustus's veterans of the battle of Philippi) (for discussion of the sources see Marchi & Sabbatini 1996, 19 – 21; Marchi 2000, 229; Marchi 2010, 40; Marchi 2014, 182 - 183). Of course, just as it is impossible to equate certain pots with certain people (see discussion in Shennan 1994; Bintliff & Sbonias 2000; Dores Cruz 2011), it is impossible to equate surface sites directly with an influx of Roman colonists. It is possible, however, to assess how the settlement organization as a whole, as reflected in site distributions, evolved over the time period for which colonization is documented to have taken place. Subsequently, we can explore how the documented trends can be related to historically known colonization events, although it will remain hard, with the current methods and data at our disposal, to tease such trends out from wider Italic or indeed Mediterranean phenomena in the present context (in general Terrenato 2007; Attema *et al.* 2010; Stek 2013, esp. 340 – 343; for Venusia see Stek 2012, 244; cf. landscapesofearlyromancolonization.com for a research design that takes this into account).

The source dataset for our analysis here consists of diachronic and hierarchical point distributions of archaeological sites recorded during field surveys (Marchi & Sabbatini 1996; Sabbatini 2001; Marchi 2010). The potential and limits of this dataset have already been discussed elsewhere (*e.g.* specifically on Venusia: Stek 2012; Casarotto *et al.* 2016; in general *e.g.* Fentress 2000 with further references).

In this paper we analyze the long-term development of the settlement landscape in two ways. First, we compared the pre-Roman settlement organization, with nucleated settlements and their possible catchment areas, to later developments in the settlement organization up to the Imperial period. In this way, we established that the pre-colonial settlement pattern played an important role in influencing the location of new settlements. Second, we tested this finding with an inductive location preference analysis, which allowed a more quantitative assessment of the correlations between settlement and cultural or natural factors. As we will see, we found a stronger effect of cultural than of natural factors on colonial period settlement locations.

6.2 DATA

The following analysis capitalizes on the legacy survey data collected during a regional, site-oriented field survey project, namely the *Forma Italiae - Ager Venusinus* project (Marchi & Sabbatini 1996; Sabbatini 2001; Marchi 2010). In addition to the identification of sites at the surface (threshold sets at five sherds per sq m) on a territory of *c.* 700 sq km, careful attention was paid during survey to those factors that could have affected the discovery of sites, such as ground visibility conditions (Marchi & Sabbatini 1996, 107; Azzena & Tascio 1996, 292 - 294; Sabbatini 2001, 59). The implications of possible distorting factors on site recovery are discussed in detail in Chapter 5; in this context, it suffices to state that there is little reason to assume that surface visibility conditions strongly biased the site patterns detected during the *Forma Italiae* survey.

In contrast to previous analyses of these legacy survey data, the data are not used here to confirm or discard

Table 6.1 Legacy survey data organized per period and per size. Archaeological sites were identified by teams of three to five surveyors spaced at five to ten m intervals, on a territory of ca. 700 sq km, using a minimum threshold of 5 sherds per sq m (see Marchi & Sabbatini 1996; Sabbatini 2001; Marchi 2010).

	Settlement size (sq m)					
	0-100	101-400	401-800	801-2000	> 2000	Tot.
Pre-Roman settlements (5th - 4th century B.C.)	100	88	34	45	22	**289**
Early colonial period settlements (3rd century B.C.)	18	34	9	18	10	**89**
Inherited settlements	7	13	6	6	7	39
New early colonial period settlements	11	21	3	12	3	50
Republican settlements (3rd - 1st century B.C.)	168	218	74	109	37	**606**
Inherited settlements	20	21	12	13	10	76
New Republican settlements	148	197	62	96	27	530
Late-Republican - Triumviral settlements (1st century B.C. - 33 A.D.)	78	138	64	93	37	**410**
Inherited settlements	22	62	25	51	27	187
New Late Republican-Triumviral settlements	56	76	39	42	10	223
Imperial settlements (1st - 4th/5th century A.D.)	144	194	78	125	53	**594**
Inherited settlements[1]	9	8	5	13	4	39
Inherited settlements	34	78	32	72	36	252
New Imperial settlements	101	108	41	40	13	303
Uncertain pre-Roman-Imperial settlements	19	12	4	7	6	**48**

[1] These sites are not occupied in the Late Republican -Triumviral period but have a Republican phase of occupation.

existing theories about Roman settlement strategies (*i.e.* a deductive approach, see Chapter 3; for another Roman context see Goodchild 2007, 180-328; 2013). Instead, the data are subjected to statistical analysis to see whether any trend in site patterning with respect to the natural and cultural environment emerges (an inductive approach) (see Kamermans & Wansleeben 1999; van Leusen & Kamermans 2005 with further references; for another Roman context see Goodchild 2007, 121-176). Only as a last step, we will consider how the detected trends compare to existing theories on Roman colonization.

In this analysis, site samples are organized both diachronically (*i.e.* per period) and hierarchically (*i.e.* per size category) (Table 6.1). Previous attempts to classify survey site types based on function (*e.g.* farm, activity area, villa, etc., see Marchi & Sabbatini 1996; Sabbatini 2001; Marchi 2010) are not considered in this paper because of the debated nature of such definitions (see the discussions in Barker & Lloyd 1991; Barker & Mattingly 1999; Bintliff & Sbonias 1999; Francovich *et al.* 2000; Rathbone 2008; Witcher 2008; Launaro 2011, 85-88; Fracchia 2013, 185 – 190; esp. for the Ager

Venusinus data see Stek 2012). By disregarding typological classifications, interpretative biases and variations in definition are limited (see van Leusen 2002, ch. 4) and site distributions can be compared on the basis of common formal attributes, such as size and period.

Another aspect must be noted at this point. We decided to consider the 3rd century B.C. settlement sample in the analyses because we are interested in assessing the settlement trends after the first colonization movement of 291 B.C.. The selection of surely-datable early colonial sites is mainly based on the presence of diagnostic and datable 3rd century B.C. black gloss pottery. Because such precisely datable finds are often not present in small sites, it is likely that the number of early colonial period sites included in this sample is an underestimation (see discussion in Chapter 3, section 3.2). It is therefore probable that several 3rd century B.C. sites are now included in the broader Republican sample (see also Marchi & Sabbatini 1996, 111 (footnote 129); Marchi 2010, 258-259). For this paper, however, we decided to ascertain whether any pattern emerges from the early colonial sample bearing in mind its small size and the likely missing evidence.

6.3 EXPLORING THE PRE-ROMAN AND ROMAN LANDSCAPES

In this first section we explore how sites in the early colonial period were placed with respect to the previous settlement organization. First, we describe this pre-Roman settlement organization in qualitative terms by comparing the site patterning to several landscape properties. Subsequently, we adopt GIS-based visualization techniques to enhance the identified relationships. We use computer-based simulation here only as a support for highlighting relationships already noted through the descriptive analysis, not as a means to find them.

The territory investigated in this paper corresponds to the north-eastern part of the modern Basilicata region, at the border with Apulia. This area is located between the southern Apennine mountains and the Apulian plain. In ancient times it was a zone of transition between different physical, but also cultural, units (*i.e.* ecotone, see Odum 1959, 278 - 280). As a matter of fact, this southern Italian territory corresponded to a "frontier" zone between the territories of Daunia (Northern Apulia), Lucania (Basilicata and southern Campania), and Samnium (central-southern Apennines), inhabited in pre-Roman times by vibrant native polities (Salmon 1967; Bottini 1980; 1982; 2016 with further references; Volpe 1990; Tagliamonte 1996; Isayev 2007; Osanna 2010; Marchi 2009; 2016 with further references). As pointed out by Olivier de Cazanove (2005), the position of this area in-between different cultural regions may well explain the choice to establish the Latin colony here, since Latin colonies were often located at frontier areas between different Italic polities (see also Torelli 1992; Marchi & Sabbatini 1996, 17 – 21; Marchi & Salvatore 1997, 5; Musti 2009; Marchi 2010, 29 – 31, 35 – 39).

From a geomorphological point of view, this "frontier" zone is characterized by the presence of two macro topographical units: the hilly landscape (an appendix of the south-eastern Apennines which extends up to the Vulture mountain) and the plateaus landscape (naturally connected to the Apulian plain through the Ofanto river). In our survey area, hills are typical of the western and south-western part, whereas large plateaus are located centrally, east-, north-, and south-eastwards (Fig. 6.1). In Archaic and pre-Roman times these two macro-regions most likely pertained to different socio-political groups. Archaeological studies suggest that the people inhabiting the major settlements of the plateau zone of this territory strongly interacted with Daunian material culture (Marchi 2010, 247-254; Bottini 2016, 10-20), while the material culture found at the settlements located in the hilly landscape demonstrates also influences of North-Lucanian traditions (emblematic is the case of the site of Ripacandida, see Bottini 1982; 2001; 2016, 33 – 42; Marchi 2010, 35, and some evidence may be attested also in Forenza and Banzi, see Marchi 2010, 181-182) (Fig. 6.1). Therefore, settlement strategies in the hilly landscape of our study area witnessed a mixture of influences from the surrounding areas, most clearly from both the Daunia and the North-Lucania regions.

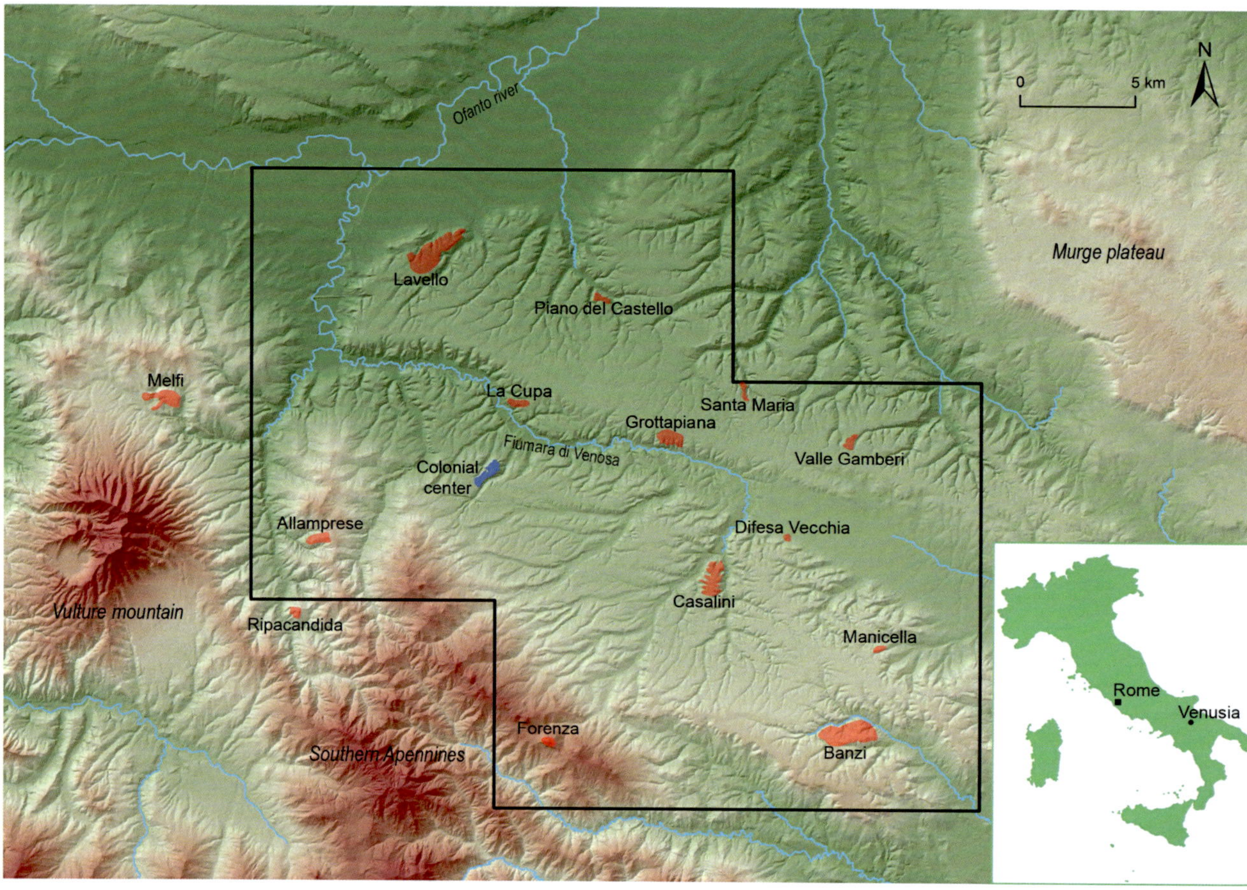

Fig. 6.1 Study area. The survey area is contoured in black. In red: major pre-Roman settlements; in blue: colonial center of Venusia. The raster base map is the shaded relief calculated from the 10-m resolution DEM named TINITALY/01 (Tarquini *et al.* 2007; 2012; Tarquini & Nannipieri 2017) and combined with an elevation color palette.

Between these two geopolitical districts (*i.e.* the hills – influences from North-Lucania and Daunia regions, and the plateaus – influences from the Daunia region), a more marginally settled area is located. Despite the identical environmental conditions (this large area consists of plateaus with the same geological and soil properties as the plateau district), it was not intensively used by pre-Roman communities (see Fig. 6.2). If, at first sight, natural factors do not justify the difference in patterning, we may consider that a cultural reason could be the cause of such a settlement vacuum in pre-Roman times. Perhaps this area represented a marginal or "frontier" zone between two different geopolitical districts (letter A in Fig. 6.2) (see discussion in De Guio *et al.* 1986), which could explain the relative lack of sites, and justify the probably different land use strategy. In the 4th and early 3rd century B.C. the whole area was affected by important socio-cultural developments that have been related to the arrival of new groups of people, indicated with the name of Samnites (Bottini 1980; Marchi & Sabbatini 1996, 99-100, 109-110; Sabbatini 2001, 57; Marchi 2016). Settlement density continued to be low in the marginal zones previously outlined, confirming the likely inheritance of land use strategies.

What is particularly interesting to us, however, is that the subsequent site patterning suggests that the decision on where to establish 3rd century settlements depended strongly on the pre-colonial landscape settlement organization (see discussion in Nuninger

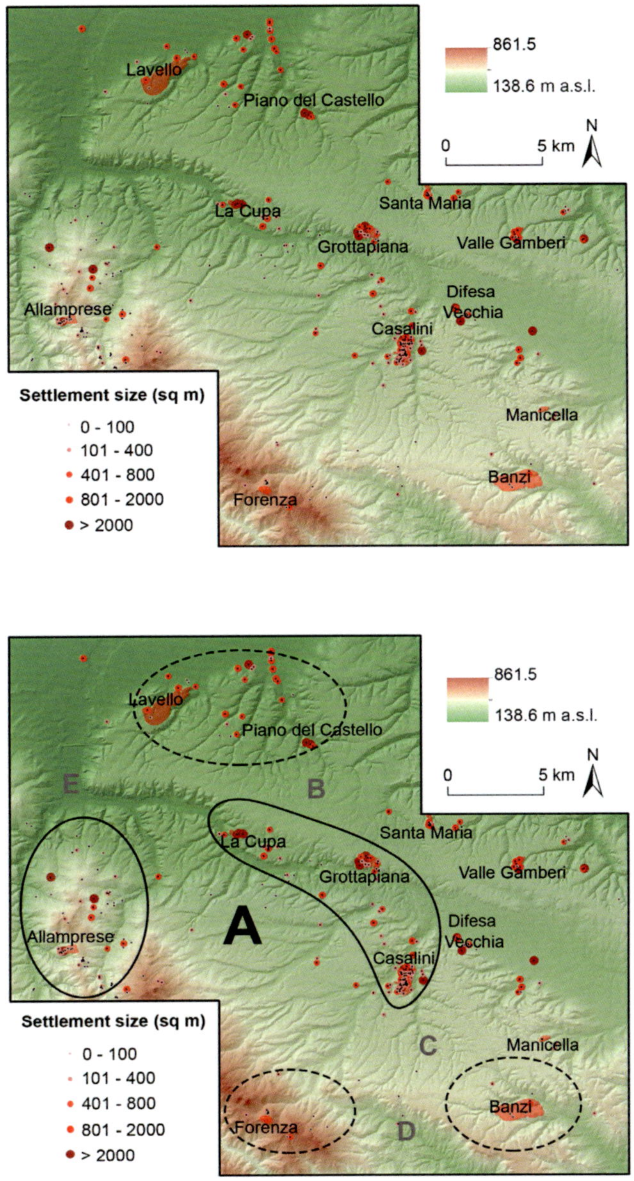

Fig. 6.2 Pre-Roman settlement distribution with the acronyms of the main nucleated settlements. The position of settlements is indicated by black dots. The extension of the red circles circumscribing these dots does not match with the scale of the map; they are used here only as symbols for the size of these settlements (see legend). Below, reconstruction of the pre-Roman territorial organization: the hilly and plateau pre-Roman districts are circled by black ellipses. Other possible districts within both the plateau and the hilly landscape are circled by dashed ellipses. The letter A indicates the position of the possible "frontier", marginal zone between the hilly and the plateau landscapes. The letters in grey indicate other zones where pre-Roman site density is remarkably low. The raster base map is the shaded relief calculated from the 10-m resolution DEM named TINITALY/01 (Tarquini *et al.* 2007; 2012; Tarquini & Nannipieri 2017) and combined with an elevation color palette.

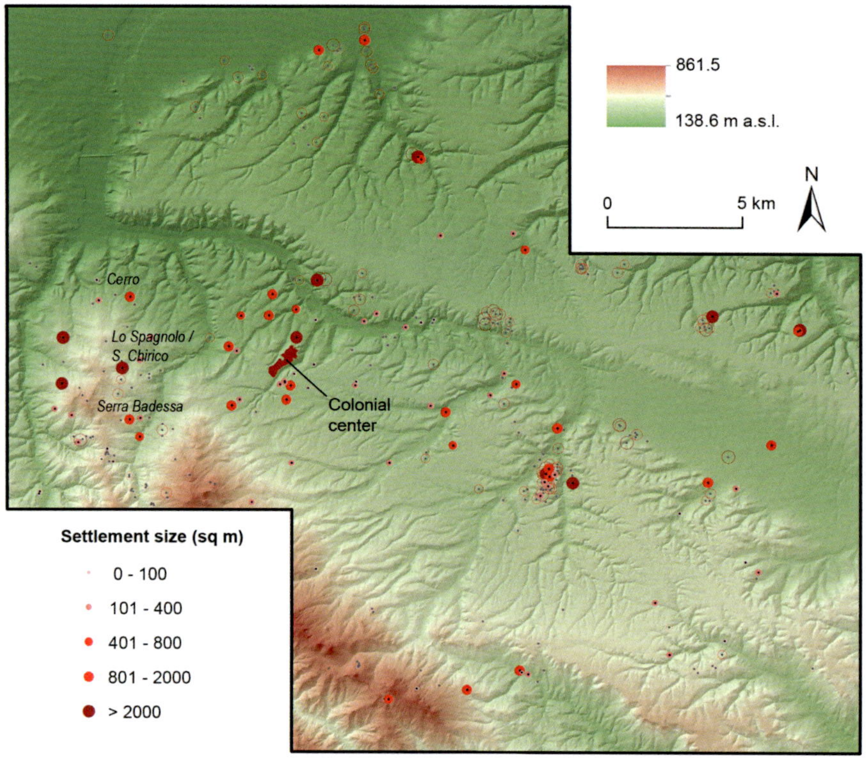

Fig. 6.3 Early colonial period settlement distribution with the indication of the position of previous pre-Roman settlements (grey dots delimitated by empty circles). The position of the early colonial period settlements is indicated by black dots. The extension of the red circles circumscribing the dots does not match with the scale of the map; they are used here only as symbols for the size of these settlements (see legend). The raster base map is the shaded relief calculated from the 10-m resolution DEM named TINITALY/01 (Tarquini *et al.* 2007; 2012; Tarquini & Nannipieri 2017) and combined with an elevation color palette.

et al. 2016). As seen in Fig. 6.2 and 6.3, 3rd century settlements were established in the previously outlined marginal and scarcely inhabited area within the native settlement system (see also Marchi 1991; Marchi & Sabbatini 1996, 111; Marchi & Salvatore 1997, 13 – 14; Marchi 2010, 249). This 3rd century settlement organization can thus be seen as an organic addition to a preexisting situation, rather than as a radical overhaul of the settlement organization, as is usually imagined. New sites were located on the plateaus around the colonial center and some others (both new sites and some that continued) were located in the hilly landscape. The period directly after the establishment of the Latin colony thus witnesses a rural infill of settlements in a previously marginally settled zone, and the majority of these sites centered around the major settlement core of Venusia. In a sense, we may therefore see the very same principle that was noted on a peninsula-wide level by de Cazanove (2005) on the local scale, namely that colonial settlement targets relatively empty and marginal areas (cf. also Marchi & Salvatore 1997, 13; Stek 2012, 244).

Another peculiarity of the 3rd century B.C. pattern must also be noted. If we look closely at its spatial distribution, it is possible to note a different settlement behavior with respect to the two geopolitical districts of the pre-Roman settlement system (see also Marchi 2004, 133). Density and distribution suggest a preference only for the hilly landscape. In the hilly landscape there is clear continuation in occupation and further rural infill from the 4th to the 3rd century B.C. (especially in the most northern part of it, in locality Lo Spagnolo,

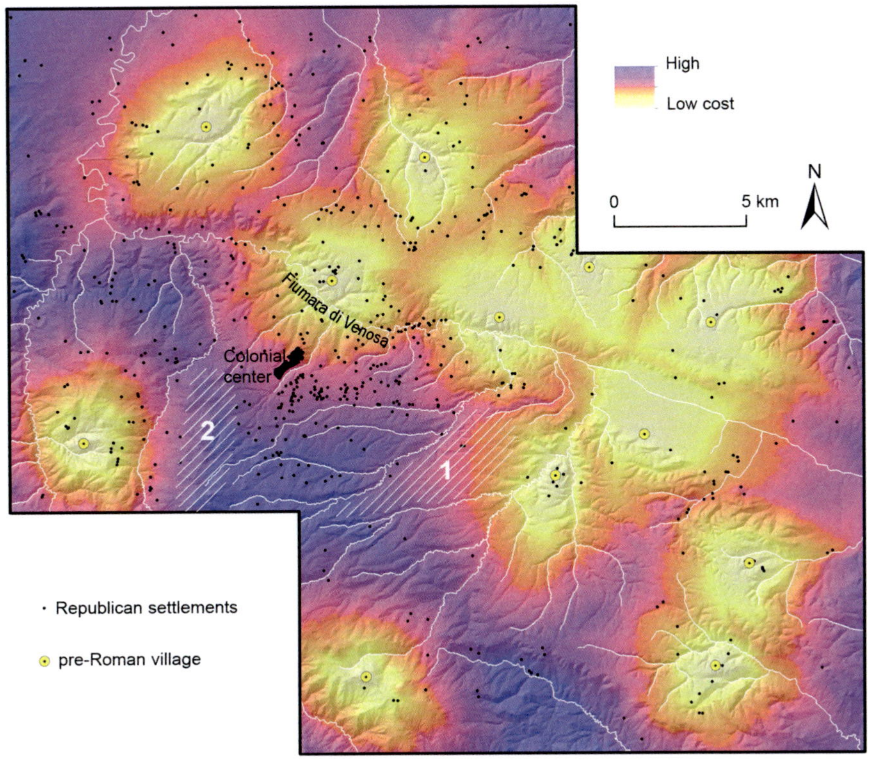

Fig. 6.4 Republican settlement distribution compared to the previous territorial organization. The centroids of the main pre-Roman nucleated villages are indicated by yellow circles. From these settlements, possible catchment areas are visualized by means of a cost analysis. The lined areas indicate the position of possible long-lasting natural or cultural "frontier" zones where site density is low from the pre-Roman to the Republican period (and even later). The raster base map is the shaded relief combined with the cost surface calculated from the 10-m resolution DEM named TINITALY/01 (Tarquini *et al.* 2007; 2012; Tarquini & Nannipieri 2017), which takes as starting points the centroids of the major pre-Roman settlements (yellow circles).

Pezza Cicoria, S. Chirico, Serra Badessa and Cerro) (see Marchi 2010, 98-117) (Fig. 6.3).

The settlement continuation in the hilly landscape has been confirmed in more recent investigations as well, through a study of the black gloss pottery collected during several surveys carried out between 2013 and 2016 in the context of the LERC project (Landscapes of Early Roman Colonization, Stek & Pelgrom 2013). With these surveys, we established that one of the major nucleated villages of the pre-Roman system, Allamprese, continued to exist in the 3rd century B.C. (Pelgrom *et al.* 2014). This situation of settlement continuity does not seem to apply to the pre-Roman plateau landscape where most villages seem to disappear fairly quickly after the establishment of the Latin colony in the area. We are currently testing this on the ground: the ongoing analyses of the material recently collected by the LERC team at the pre-Roman nucleated settlements of the plateau landscape (*e.g.* Casalini, Piano del Castello) will allow more accuracy in the future (see Pelgrom *et al.* 2016). From the dataset at our disposal, however, it appears that the hilly landscape continued to be occupied in the 3rd century B.C., and seems to be more integrated into the early colonial settlement system than the plateaus landscape located to the north, east, and south-east of the survey sample area (see also Marchi and Sabbatini 1996, 114 – 115).

The impression given by the 3rd century settlement sample in the plateau landscape, thus, is that the new settlements initially may have adapted themselves to a preexisting situation by occupying the space still available. Also, more generally in the Republican period, the site patterning continues to display the characteristics of an adaptive rural infill both in the same preferred areas of the early colonial phase (hills and central plateaus) and in the marginal and less-densely settled zones of the pre-Roman settlement organization.

The Republican period rural infill, indeed, does not affect all landscape districts equally, but seems to concentrate in certain niches. Some clear trends in the use of space can be detected. For instance, certain areas remain blank from the pre-Roman to the Republican periods, suggesting perhaps the presence of natural or cultural "frontier" zones (*e.g.* woods, sacred places) that were respected by not being settled over time (see Fig. 6.4 for a possible reconstruction of some of these "frontiers"). On the other hand, in other previously unoccupied niches we do see an infill of sites occurring in the Republican period. Moreover, the role played by the colonial urban center remains clear: the rural infill clusters around and expands from the town over time.

This clustering of Republican period sites both in previous marginal niches and close to the colonial town is displayed in Fig. 6.4. This image represents a simulation that models the pre-Roman network of major nucleated settlements, their territorial catchments, and the marginal zones and niches in between them. We conducted this analysis in TerrSet GIS by using the module 'Varcost' (Eastman 2016). Territorial catchments (Vita-Finzi & Higgs 1970) were calculated on the basis of an anisotropic cost surface that takes into account as obstacles to human movement both the morphology of the terrain (altitude, slope and aspect) and the position of the main not-navigable rivers[27]. The extension of each catchment area terminates when the same cost value calculated from the neighbor settlements is met (for more sophisticated ways of simulating settlement territories see Renfrew & Level 1979; De Guio 1988-1989; De Guio & Secco 1988; Wheatley & Gillings 2002, 151-159; Conolly & Lake 2006, 208-225; Ducke & Kroefges 2008).

Fig. 6.4 illustrates that the colonial urban center, and the settlements clustered around it, are located in the largest purple/pink halo corresponding to the marginal zone between the two geopolitical districts of the pre-Roman settlement system. Interestingly, the other Republican settlements tend to cluster in smaller niches, precisely at the margins of the catchment areas of the pre-Roman nucleated settlements.

We also noted that the rural infill by Republican settlements, but also by the subsequent Triumviral ones, does not affect the entire survey sample area with the same density. In fact, the plateaus located south and east in the survey area are much less densely occupied. The Republican and Triumviral samples clearly tend to concentrate in the central plateaus (around the town), in the northern plateaus (on the other side of the *Fiumara di Venosa* valley), and west in the survey area. The dividing line between these two different trends in regional pattern and density can be positioned on a large plateau that we previously pointed out as one possible long-lasting "frontier" zone (see number 1 in Fig. 6.4).

This situation changes in the Imperial period, when sites are more homogeneously and regularly located across the entire survey area (Figs. 6.6-6.7) (see also Marchi 2004, 139), and also occupy those southern and eastern parts of the survey sample area that witnessed no remarkable settlement in the previous Roman periods[28]. Therefore, for Imperial rural settlement strategies, the pre-existing landscape organization seems to play a minor role in location preferences.

27 The elevation dataset used to calculate the altitude, slope, aspect and some other variables (see Fig. 6.5 and Figs. 2.1-5; 3.1-7; 3.10; 4.1-6; 5.1-3; 5.6; 6.1-9; B.1) is the 10 m-resolution DEM named TINITALY/01 (Tarquini *et al.* 2007; 2012; Tarquini & Nannipieri 2017). Permission for download and use of this dataset was released in May and June 2014 (for further information see http://tinitaly.pi.ingv.it/).
Rivers and streams were extracted from the hydrological system of Regione Basilicata. This data was kindly provided by Regione Basilicata in June 2013. Data concerning the hydrography of this territory can be found in the catalogue of the Geoportale of Basilicata (RSDI): http://rsdi.regione.basilicata.it For the territory outside Basilicata, rivers and main streams were digitalized manually on the basis of topographic maps.

28 This is in line with the taking over of villa estates consisting of large Imperial settlements and other satellite productive settlements from the 1st century A.D. onwards (see also Torelli 1991, 23; Marchi 2004, 129-130; Marchi 2010, 260) which were spread homogeneously to exploit all different ecological sectors of this landscape (see discussion in White 1970, 19-20; Gabba 1977, 269-284; Torelli 1990; Carandini 1994; Capogrossi Colognesi 2002; Gualtieri 2002; Marchi 2010, 273, 281; Terrenato 2012).

6.4 LOCATION PREFERENCE ANALYSIS

To further test the trends highlighted before, and to see whether there are other correlations which can help explain the resulting patterns, we now describe the outcomes of the inductive location preference analysis. The technical explanation and discussion of the methods and statistical tests used for carrying out such an analysis is provided elsewhere (Casarotto 2017). Here, we only list the detected location preferences per period and size category and highlight whether these preferences change through time. Remarkable changes in location preferences may indicate significant changes in settlement strategies, which may, at least theoretically, be linked to the historically documented colonization movements. Thus, these changes may provide a measurement for the impact of Roman colonization.

We tested for the presence of spatial correlations between site patterning and several environmental and cultural characteristics of the landscape[29]. Specifically, the distribution of diachronic and hierarchical site samples was compared to altitude, slope, aspect, soil, location of dominant positions in the landscape (*i.e.* ridges and peaks), distance from water, distance from the city and from the major Roman roads (Fig. 6.5 A and B). To assess whether the impact of Roman colonization is visible starting from the early colonial sub-phase, two comparisons with these variables were carried out. The first is a macro-comparison between the generally datable pre-Roman, Republican, and Imperial settlements aimed at testing whether the settlement rationale drastically changed from the pre-Roman to the Republican period, or only later in time. The second comparison is a micro-comparison, meaning that we zoomed in on the early colonial sub-phase to evaluate whether the most significant divergences in distribution with the pre-Roman phase emerged in the 3rd century BC.

6.4.1 FIRST COMPARISON: PRE-ROMAN – REPUBLICAN – IMPERIAL SETTLEMENTS

The highest variability in location preferences is exhibited by the smallest site categories (0-100 and 101-400 sq m settlements) that have the most typical and diverging distribution in the different periods with respects to the variables considered in this analysis. On the other hand, larger settlement types have, in general, similar location preferences, independent of the period. The similar preferences attested by the largest settlements are mostly dependent on the fact that there exists a direct relationship between settlement size and survival rate: larger settlements very often attest continual occupation in several phases whereas the smaller site types have a shorter life span (usually one phase of occupation) and attest a higher variability in location preferences.

When analyzing the totality of sites per period, we noted a divergent preference of ecological districts in the pre-Roman and Republican phase. These different land units can be outlined by different altitude bands and soil conditions. The problem we want to tackle here is whether this difference in preference was possibly dictated by the more favorable soil types and geomorphological conditions characterizing the different land units. Pre-Roman settlements tend to concentrate on the 4th and 5th altitude band (and avoid the 1st one), and they seems to be especially attracted by soil unit 6.3 and 6.4 in the hilly landscape (see Table 6.2). The Republican settlement system, instead, preferentially clusters in the plateau landscape (quite significantly in the 3rd altitude band), which is mainly characterized by the presence of soil unit 11.1. As suggested in the previous descriptive analysis, this clustering may be explained as a result of cultural constraints rather than as a function of natural factors. Here, we want to further test this assertion.

29 For this aim we used distribution graphs and statistical tests (see Appendix B). Specifically, the Chi-squared test (Siegel 1956, 42-47; Shennan 1988, 65 -70), the Kolmogorov-Smirnov test (Siegel 1956, 47-52; Wheatley & Gillings 2002, 136-142) and the Attwell-Fletcher test (Attwell & Fletcher 1985; 1987) were applied to highlight possible significant correlations in the dataset (Kvamme 1990). A significance level (α) of 0.05 was selected for these analyses. For an extensive explanation of the method and lists of the results, reference is made to Casarotto (2017) and to Appendix B of this book.

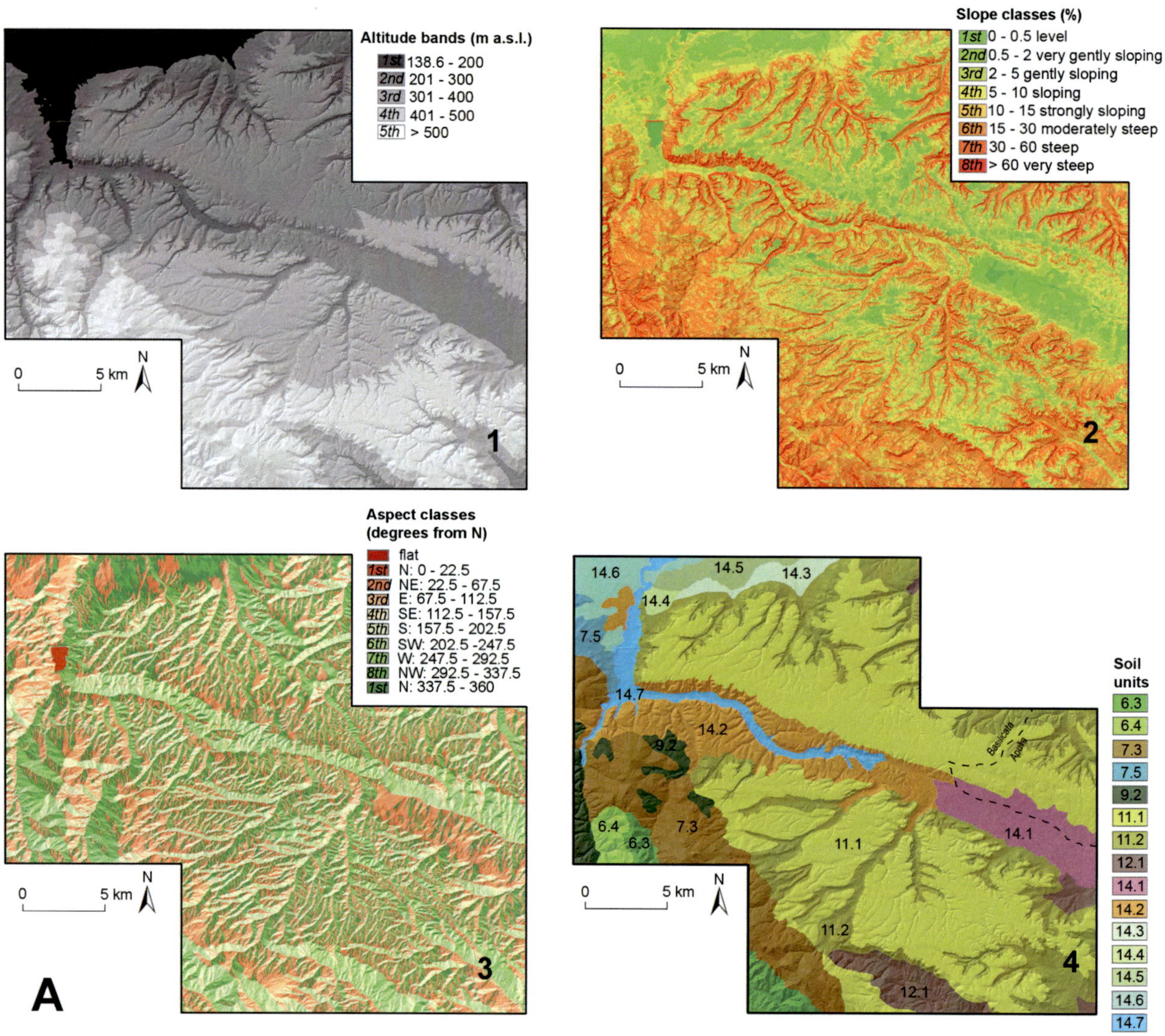

Fig. 6.5 A Variables considered in the inductive location preference analysis calculated in IDRISI GIS (Eastman 2012), ArcGIS 10.2.2 and LandSerf GIS (Wood 2009). Detailed information on how these variables were calculated is in Casarotto (2017). A: 1) Altitude (based on the 10 m-resolution DEM named TINITALY/01, Tarquini *et al.* 2007; 2012; Tarquini & Nannipieri 2017); 2); Slope (calculated from the 10 m-resolution DEM named TINITALY/01, Tarquini *et al.* 2007; 2012; Tarquini & Nannipieri 2017). The classification in slope classes is based on FAO 2006 (p. 12); 3) Aspect (calculated from the 10 m-resolution DEM named TINITALY/01, Tarquini *et al.* 2007; 2012; Tarquini & Nannipieri 2017). The classification in aspect classes is based on Esri 2014; 4) Soil. The base map for the territory within the administrative borders of the Basilicata region is the soil map of the Regione Basilicata (1: 250,000) (Ufficio Produzioni Vegetali e Silvicoltura Produttiva - Dipartimento Agricoltura, Sviluppo Rurale, Economia Montana - Regione Basilicata). Outside this territory soil properties were reconstructed (for further information see footnote 30, p. 116).

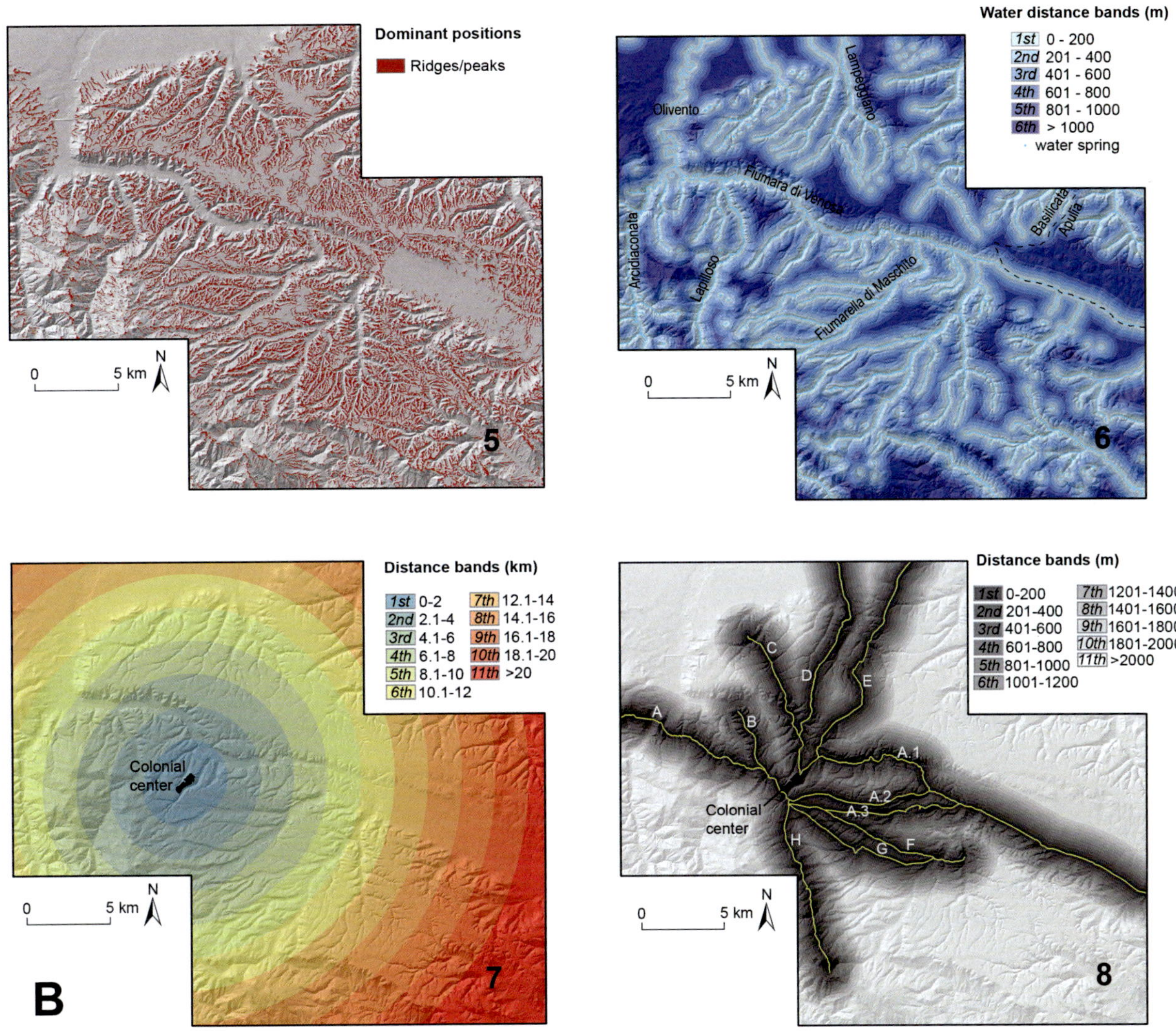

Fig. 6.5 B 5) Location of dominant positions in the landscape (calculated from the 10 m-resolution DEM named TINITALY/01, Tarquini *et al.* 2007; 2012; Tarquini & Nannipieri 2017); 6) Distance from water sources (main rivers, main streams, and perennial water springs) (see also footnote 27, p. 112); 7) Distance from the town of Venusia; 8) Distance from (Roman) roads.

Based on the soil map of the area (Regione Basilicata, 1:250,000)[30], the detected location preferences cannot be justified by the presence of more favorable soils[31]. The sandy conglomeratic soils (unit 11.1), on which a majority of Republican sites are found, are not particularly suited for arable farming (but see further discussion in Van Joolen 2003) and, due to the abundant presence of conglomeratic pebbles coming from the bed rock, are also quite difficult to work with basic tools (Frayn 1979; Spurr 1986; Goodchild 2007, 147). Interestingly, even the small Republican settlements (*i.e.* possibly mononuclear colonists' farms, Marchi & Sabbatini 1996, 111-115) do not exhibit significant correlations with those land units that, for their good properties (high/medium fertility and high/medium workability, see Table 6.2), might have been more suitable for carrying out subsistence agriculture (tracts of these 'unjustified' empty spaces in the Republican settlement pattern have been already noted by Marchi & Sabbatini 1996, 114; Sabbatini 2001, 71). It seems that the choice of preferentially settling this district in the Republican period was dictated by other constraints, such as indeed cultural ones, as we will discuss in more detail further below.

In addition to the influence of the previous territorial organization, another important cultural element is the colonial urban center. A significant preference to stay close to the main colonial settlement is clear for the Republican settlements (see also Marchi & Sabbatini 1996, 112 – 114; Marchi 2004, 133). The first two distance bands (0 – 4 km) have a significantly high density of sites, independently from the size category (only the largest Republican settlements - > 2000 sq m - have a more scattered distribution with respect to the center). Additionally, (Roman) roads departing from or crossing the center (for a reconstruction see Fig. 6.9, p. 121) correlate with settlement density: again, a significantly high density of Republican sites is attested in the first two distance bands (0-400 m) but the largest Republican settlements are more spread out across the territory. The colonial urban center displays a correlation with later settlements as well (*i.e.* Late Republican - Triumviral and Imperial settlements), but this influence decreases significantly especially in the Imperial period. Imperial settlements are more homogeneously distributed with respect to the distance from the center (see also Marchi & Sabbatini 1996, 117 – 123; Sabbatini 2001, 72 – 75; Marchi 2004, 139; Marchi 2010, 43). A weaker correlation is observed for the distance from a (Roman) road as well (cf. Marchi 2010, 281).

Previously we noted a lack of interest in settling the best land units in terms of fertility and workability during the Republican period. This situation seems to change in the Triumviral but, especially, in the Imperial period. As a matter of fact, at this moment in time a significantly high concentration of settlements is attested in correspondence with highly fertile alluvial and volcanic soils (units 14.2 and 9.2). Interestingly, the preference for volcanic soils is also displayed by the smallest Imperial sites. Possibly, this preference can be related to strategic or specialized land-use practices adopted during this period of time. Unfortunately, based on the data available it is impossible to be more precise about which type of cultures were more likely to have been cultivated here (for hypothetical reconstructions in Roman contexts see the land evaluation analyses by Van Joolen 2003; Goodchild 2007). An indication of the most suitable type of crops may, however, be deduced from the modern land use: today, in these portions of the landscape, olive groves and vineyards are grown because of the optimal soil characteristics for these crops.

To conclude, the altitude and soil variables are the environmental factors exhibiting the clearest

30 Ufficio Produzioni Vegetali e Silvicoltura Produttiva - Dipartimento Agricoltura, Sviluppo Rurale, Economia Montana - Regione Basilicata. Data and legend can be found here: http://www.basilicatanet.it/suoli/index.htm (credits: http://www.basilicatanet.it/suoli/credits.htm) and in the catalogue of the Geoportale of Basilicata (RSDI): http://rsdi.regione.basilicata.it. The shapefile of the soil map of Basilicata was kindly provided by Regione Basilicata in May 2013.
The outmost east corner of the survey sample area belongs to the Apulia Region. The soil information for this small zone was inferred by the first author (AC) on the basis of the physiographic and geological conditions. The geological maps of this area (Carta geologica d'Italia 1: 500,000 - Geoportale Nazionale - Ministero dell'Ambiente e della Tutela del Territorio e del Mare, and Carta geologica d'Italia 1: 100,000 – Foglio 188, Servizio Geologico d'Italia) were controlled to map the soil units in this zone: since the geomorphological and geological characteristics of this area are the same of adjacent soil units (*i.e.* 11.1, 11.2 and 14.1; see also the Carta geologica d'Italia 1: 100,000 – Foglio 175 and 187, Servizio Geologico d'Italia), this small portion of the survey area was classified accordingly, using these soil types (see Figure 6.5 A, p. 114).

31 It is important to underline, however, that the present natural conditions and the present distribution of soil types may be different to those which existed in the past. Also, the scale of the soil map might have influenced our analysis in certain zones of the landscape.

Table 6.2 Soil units. This is a basic classification based on the information provided by the legend of the soil map of Regione Basilicata. For more detailed descriptions of soils and soil properties see http://www.basilicatanet.it/suoli/carta2.htm ; http://www.basilicatanet.it/suoli/province.htm ; https://rsdi.regione.basilicata.it. For these land units a qualitative evaluation of the suitability for general agricultural purposes (*i.e.* plant growth) is proposed (see last column). The productive potential (*e.g.* low, medium, high) of each land unit is established on the basis of two important qualities of the soil (see also Vink 1975, 196 – 208), namely fertility (here depending on the availability of nutrients and minerals, and the drainage status of the soil) and workability (here depending on slope and stoniness qualities, White 1970; Frayn 1979; Spurr 1986). In principle, abundance of plant nutrients and minerals along with a good drainage are typical of fertile soils; flat to gently sloping surfaces with scarce presence of stones are typical of easily workable soils. The land qualities from which fertility and workability are inferred (cf. supra) have been estimated on the basis of the information provided in Vink 1975, Kamermans 2000, FAO 2014, and in the legend of the soil map. In addition to that, for several of these units the land qualities related to workability could also be assessed directly, in the field, during recent archaeological field surveys, in which surveyors recorded systematically both slope and stoniness conditions of the fields (LERC survey campaigns 2013 – 2016, see Pelgrom *et al.* 2014, Pelgrom & Tetteroo 2015; https://landscapesofearlyromancolonization.com/).

Soil unit	Area (sq km)	%	Landscape type	Topography	Geology	Soil type WRB 98	Modern land use	Fertility	Workability with basic tools	Suitability for agriculture
6.3	13.2	1.9	Mountains	Moderately steep to very steep	Quartz sandstones with thin layers of clay rocks	Eutric Cambisols; Endogleyi-Luvic Phaeozems	Mainly forest	Medium/Low	Low	Low
6.4	6.7	1	Mountains	Gently sloping to steep	Sandstones and marls	Eutric Cambisols	Forest and pasture	Medium	Medium	Medium
7.3	67.3	9.7	Hills	Undulating	Clayey slate rocks and marls	Luvi-Vertic Phaeozems; Calcaric Regosols	Arable	Medium/Low	Medium	Medium/Low
7.5	7.4	1.1	Surfaces connected the hills with the alluvial landscape	Flat to gently sloping	Clayey marls	Luvi-Calcic Kastanozems	Arable	Medium	Medium	Medium
9.2	15.8	2.3	Hills	Gently sloping to moderately steep	Pyroclastic colluvial deposits	Luvic Phaeozems; Eutric Cambisols; Dystri-Andic Cambisols	Mainly viticulture and olive orchards alternate to forest and pasture	High	Medium	Medium/High
11.1	280.0	40.2	High plateaus (ancient Pleistocene surfaces)	Flat to gently sloping	Sands and Pleistocene conglomeratic deposits	Luvi-Vertic Kastanozems; Luvic Kastanozems; Calcic Vertisols	Arable	Medium	Medium/Low	Medium/Low
11.2	136.0	19.5	Slopes of the higher plateaus	Gently sloping to steep	Sands and Pleistocene conglomeratic deposits	Luvic Kastanozems; Eutric Cambisols; Calcari-Arenic Regosols	Arable	Medium/Low	Low	Low
12.1	23.3	3.4	Hills	Undulating	Clayey and silty marine deposits, mainly marls	Hyposodic Vertisols; Luvi-Vertic Kastanozems	Arable	Medium	Medium	Medium

14.1	29.0	4.2	Alluvial plain	Flat	Fluvio-lacustrine deposits (with pyroclastic material)	Pelli-Calcic Vertisols	Arable	High	High	High
14.2	61.5	8.8	Low plateaus (fluvio-lacustrine terrace)	Flat	Fluvio-lacustrine deposits (with pyroclastic material)	Luvi-Vertic Phaeozems; Calcic Vertisols	Arable and pasture	High	High	High
14.3	8.9	1.3	Surfaces connected the plateaus with the alluvial landscape	Flat to gently sloping	Alluvial deposits and colluvial deposits with clayey and sandy granulometry	Eutri-Vertic Cambisols	Arable	Medium	High	Medium/High
14.4	2.5	0.4	Alluvial terraced conoids	Flat to gently sloping	Sandy and clayey deposits	Calcic Luvisols	Arable	High	High	High
14.5	9.7	1.4	Alluvial terraces	Flat to gently sloping	Sandy, clayey and silty deposits	Petric Calcisols; Eutri-Fluvic Cambisols	Arable	Medium	Medium	Medium
14.6	15.4	2.2	Alluvial terraces	Flat to gently sloping	Higher clayey and silty deposits, lower sandy and gravelly deposits	Luvic Phaeozems; Haplic Calcisols; Eutric Vertisols	Arable	Medium	High	Medium/High
14.7	19.1	2.7	Valley floors	Flat	Sandy and stony deposits	Eutri-Fluvic Cambisols; Calcaric Phaeozems; Calcari-Arenic Regosols	Arable	Medium/Low	Medium	Medium

difference in location preferences between pre-Roman, Republican and Imperial periods (Figs. 6.6 and 6.7). The other natural factors (*i.e.* slope, aspect, preference for dominant positions, and distance from water) affect settlement distribution quite uniformly across the various periods. Indeed, all settlement samples have similar distributions with respect to slope and aspect conditions of the terrain (see also Figs. B.2, B.7 and B.8 in Appendix B). However, since a high number of pre-Roman settlements is located on the hills, we noticed some differences in slope values, which are however not statistically significant.

Dominant positions (ridges and peaks) attract settlement interests in the pre-Roman, Republican, and Imperial periods (but the larger settlements of these periods do not seem to be particularly attracted by such locations, probably because of their large extent which required open spaces such as plateaus). The distance from main rivers, main streams, and perennial water springs seems to be a factor that has been taken into consideration when deciding where to locate a settlement. As a general trend, sites tend to be located at a certain distance from rivers (probably due to the risk of flooding at the nearby locations) but close enough to reach them easily. Settlements thus tend to avoid the farthest distance bands and to favor the second distance band (200 to 400 m from a water source). This is especially highlighted by both the smallest Republican settlements (0-100 sq m) and by the small pre-Roman settlements (101-400 sq m). We conclude that a regular water supply is an important factor for settlement purposes in all historical periods considered in this analysis.

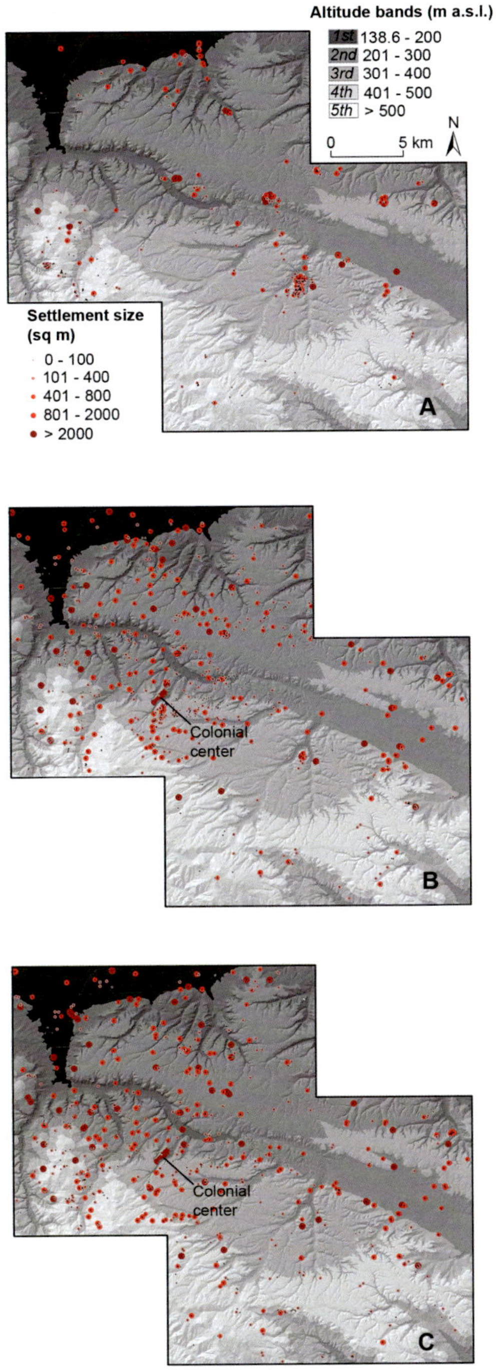

Fig. 6.6 Pre-Roman (A), Republican (B) and Imperial (C) settlement distributions compared to the altitude variable. The position of the settlements is indicated by black dots. The extension of the red circles circumscribing the dots does not match with the scale of the map; they are used here only as symbols for the size of these settlements (see legend). The raster base map is the shaded relief calculated from the 10-m resolution DEM named TINITALY/01 (Tarquini *et al.* 2007; 2012; Tarquini & Nannipieri 2017).

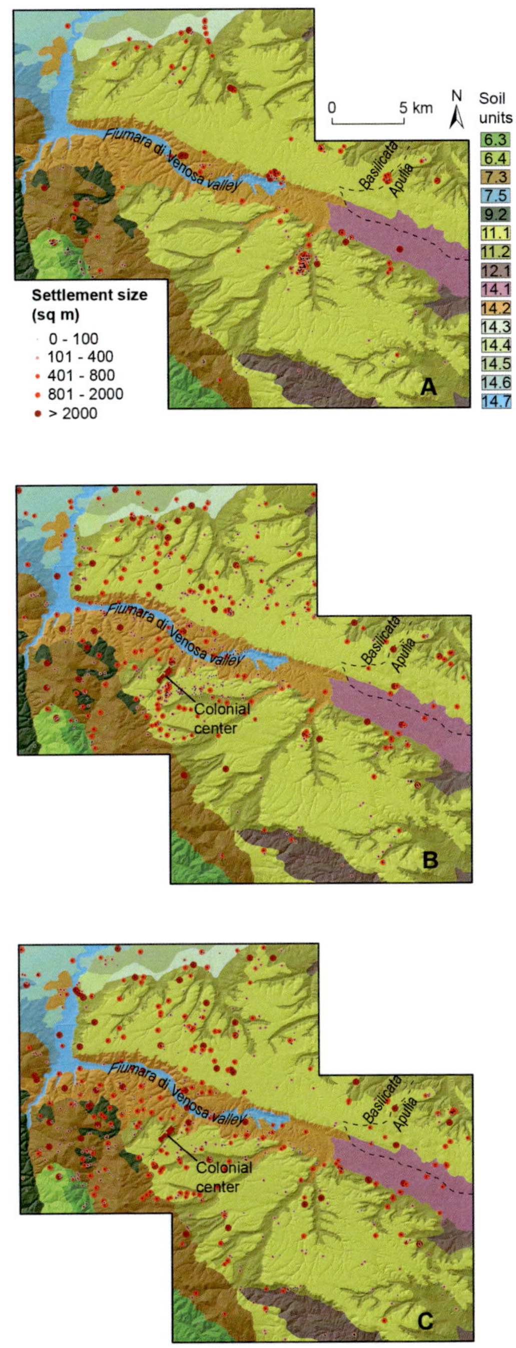

Fig. 6.7 Pre-Roman (A), Republican (B) and Imperial (C) settlement distributions compared to the soil variable. The position of the settlements is indicated by black dots. The extension of the red circles circumscribing the dots does not match with the scale of the map; they are used here only as symbols for the size of these settlements (see legend). The base map for the territory within the administrative borders of the Basilicata region is the soil map of the Regione Basilicata (1: 250,000) (Ufficio Produzioni Vegetali e Silvicoltura ▶

◀ Produttiva - Dipartimento Agricoltura, Sviluppo Rurale, Economia Montana - Regione Basilicata). Outside this territory, soil properties were reconstructed, for further information see footnote 30, p. 116. The raster base map is the shaded relief calculated from the 10-m resolution DEM named TINITALY/01 (Tarquini *et al.* 2007; 2012; Tarquini & Nannipieri 2017).

6.4.2 SECOND COMPARISON: PRE-ROMAN – NEW SETTLEMENTS IN THE EARLY COLONIAL PERIOD – EARLY COLONIAL PERIOD SETTLEMENTS

In this section, we aim to point out early colonial period location preferences to assess significant differences (if any) with the pre-Roman settlement distribution. No clear differences between the location preferences of pre-Roman and early colonial period settlements were detected with the environmental variables considered in this analysis (*e.g.* altitude, slope, distance from water sources). The only clear change in settlement strategies from the pre-Roman to the early colonial period can be related to the colonial central place. There is a clear tendency by new early colonial period settlements to cluster around the center of Venusia (a significant preference is attested for the first distance band: 0-2 km from the center). New early colonial period sites may be attracted by roads as well (see Fig. B.5 in Appendix B, p. 142). Again, we observed that cultural variables play the most influential role in settlement strategies during the early colonial period (see Figs. 6.8 and 6.9).

Cultural variables thus seem key factors not only on the peninsular scale (De Cazanove 2005; Stek 2018) and regional scale (Marchi & Sabbatini 1996, 115; Stek 2012), but also on the local scale, where the colonial center and, to a lesser extent roads, influence the internal logistical structure of the early colonial period settlement pattern. Therefore, we conclude that, contrary to what is conventionally suggested, the pre-colonial settlement organization and the colonial center are the most influential factors in early colonial period settlement strategies, instead of soil and other environmental conditions.

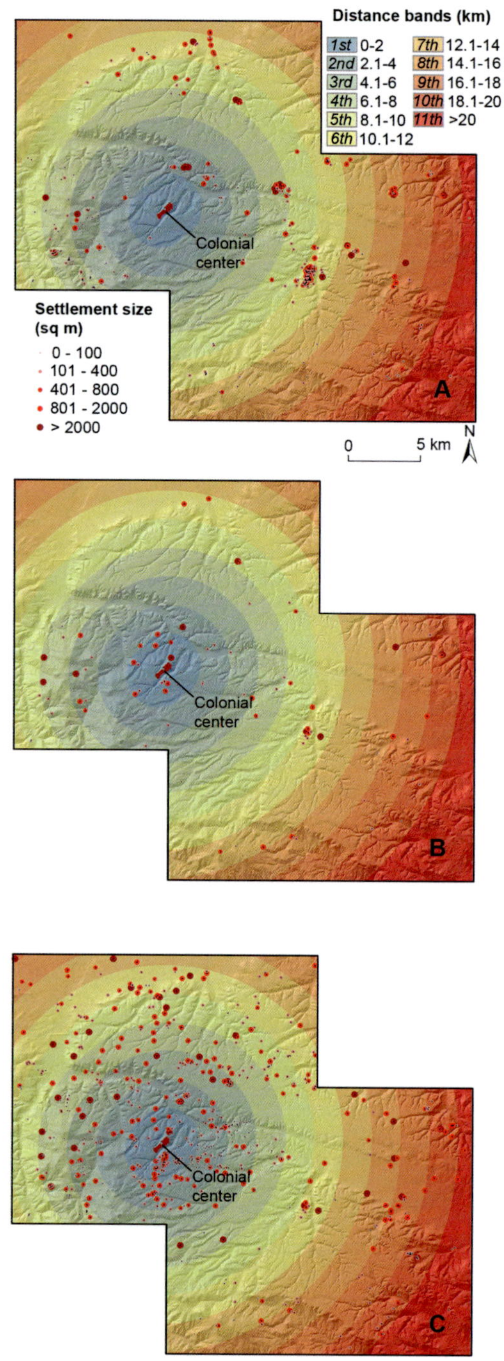

Fig. 6.8 Pre-Roman (A), Early Colonial (B) and Republican (C) settlement distributions compared to the distance from the colonial center. The position of the settlements is indicated by black dots. The extension of the red circles circumscribing the dots does not match with the scale of the map; they are used here only as symbols for the size of these settlements (see legend). The raster base map is the shaded relief calculated from the 10-m resolution DEM named TINITALY/01 (Tarquini *et al.* 2007; 2012; Tarquini & Nannipieri 2017).

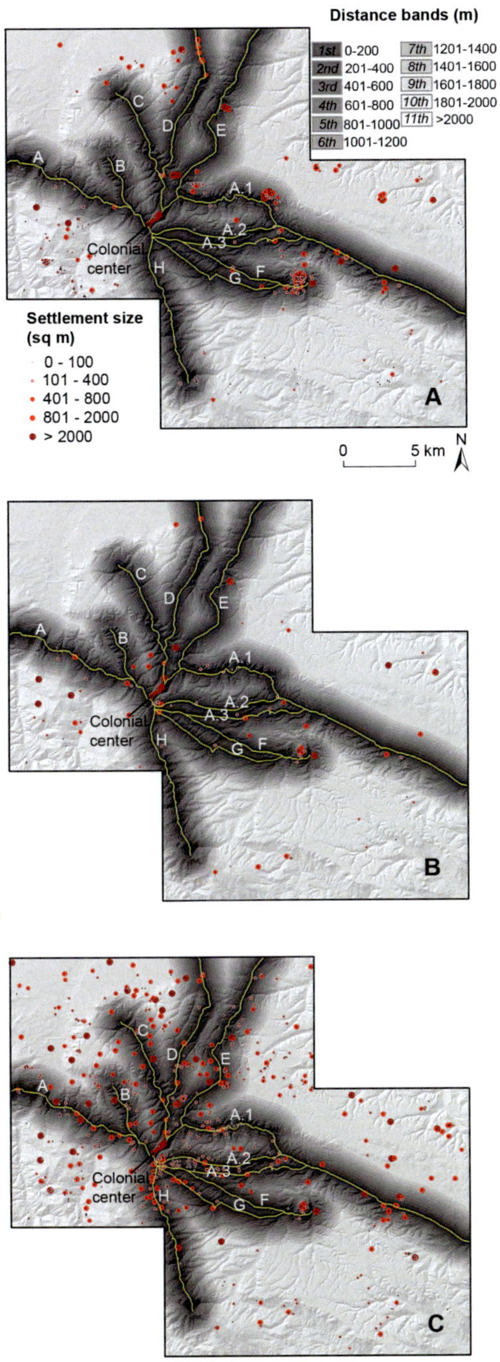

Fig. 6.9 Pre-Roman (A), Early Colonial (B) and Republican (C) settlement distributions compared to the distance from a Roman road. The routes have been digitalized on the basis of the information and maps reported in Salvatore 1984 (pp. 17 - 21), Marchi & Sabbatini 1996 (pp. 125 - 127), Sabbatini 2001 (pp. 78 - 80) and Marchi 2010 (pp. 279 - 285). A: via Appia; A.1: alternative route of the via Appia (Marchi & Sabbatini 1996, 125); A.2: segment of the via Appia (Marchi & Sabbatini 1996, 125 - 127, see also Marchi 2010, 279 - 285) ; A.3: segment of the via Appia (Vinson 1979; Marchi & Sabbatini 1996, 125 - 127); B: via Venusia - Herdonias (Salvatore 1984, 17 - 21; Marchi & Sabbatini 1996, 125 - 127); C: via Venusia - Forentum (Marchi & Sabbatini 1996, 125 - 127); D: road parallel to the Lampeggiano river (Sabbatini 2001, 78 - 80); E: via Venusia - Canusium (Sabbatini 2001, 78 - 80). F and G: via Venusia - Bantia (Buck 1981; Marchi & Sabbatini 1996, 125 - 127); H: via Herculia (Buck 1971; Marchi & Sabbatini 1996, 125 - 127). The raster base map is the shaded relief calculated from the 10-m resolution DEM named TINITALY/01 (Tarquini *et al.* 2007; 2012; Tarquini & Nannipieri 2017).

6.5 CONCLUSIONS

In order to understand the settlement development of the area around the town of Venusia, we carried out two different analyses to investigate how site patterning evolved over time. In both analyses we focused particularly on those historical moments that, according to the literary sources, witnessed the arrival of groups of Roman settlers who colonized this territory. Based on the results from the first descriptive analysis, we therefore conclude that the pre-existing territorial organization of the pre-Roman settlement system determined to a large extent the choice of the area where the first 3rd century colony was established. The most influencing factors in early colonial period settlement strategies are therefore not to be sought in favorable conditions of the natural environment, but rather in cultural constraints. The newly developing settlement organization complemented, rather than replaced the existing territorial organization, by adaptively filling in the relatively scarcely settled portions of the landscape, corresponding to the marginal niches at the borders of the catchment areas of pre-Roman nucleated settlements and the large empty zone consisting of plateaus located centrally in the survey sample area. The weak effect of natural factors (*e.g.* soil) on early colonial period settlement developments has also been confirmed by means of the second inductive analysis of location preferences. In addition, through this analysis we established that the internal structuration of the settlement infill also depended mostly on cultural factors. Most evidently, the distance from the main colonial settlement-core had an important role to play in rural settlement location preferences and, to a lesser extent, the distance from roads as well.

Contrary to what has been proposed in previous debates on Roman colonization, our analyses have made clear that the pre-existing settlement organization had a determining role in the development of colonial period settlement strategies. Of course, it is not possible from survey evidence to link the new settlements directly to colonization. But the gradual and adaptive settlement development in the colonial landscape we saw, certainly challenges the conventional notion of a radical break with previous settlement organization and its settlers (contra *e.g.* Settis 1984; see also the discussion in Terrenato 2001; 2007; Bradley 2006; van Dommelen & Terrenato 2007; Roppa & van Dommelen 2012; Stek *et al.* 2015; Sewell 2016; Vermeulen 2017).

This more adaptive and organically growing settlement scenario might also explain the presence of localized densities of Republican settlements at the margins of the pre-Roman settlement system we noted in previous work (for Aesernia see Stek *et al.* 2015; for a comparison between Venusia, Aesernia and Cosa see Casarotto *et al.* 2016, *i.e.* Chapter 3 of this book). Due to the high isotropy in natural conditions of the plateau area, this cultural trend is particularly evident in the plateau landscape (see Fig. 6.4, p. 111). In the hilly landscape such a clustering trend at the margins is not immediately visible, because the high variability in morphological conditions characterizing this unit hampers a clear discrimination of marginally settled niches. However, when we look more carefully at the patterning, we may also recognize the same trend here. Indeed, we noted that Republican concentrations of sites tend to cluster in previously less densely settled zones, just in front of prior nucleated pre-Roman settlements (*e.g.* in Pezza Cicoria, Serra Badessa, Lo Spagnolo) (see Marchi 2010, 253-255 for a different explanation). For the Republican period, therefore, we conclude that the dominant type of regional pattern consists of a balanced rural infill of sites clustered in few scattered zones only marginally occupied in the previous period. Hypothetically, if we assume all Republican period settlements were established in the early colonial phase by people of Roman origin, data patterning suggests that they probably adapted to a preexisting situation by preferentially occupying the available space, regardless of its natural properties. The 3rd century as well as the 2nd century B.C. waves of Roman colonization, apparently, did not trigger a complete restructuring of the territorial organization, as is usually suggested.

In the late 1st century B.C., natural factors started to play a more important role in allocation strategies. In the Triumviral period, indeed, the only remarkable change in location preferences we observed from the Republican phase is a preference for more fertile soils. However, in this period cultural constraints also still seem to be more influential with respect to territorial organization. As a matter of fact, a further rural infill in correspondence to the same districts favored by previous Republican settlements could be noted. This changed quite drastically only in the Imperial period. As Marchi and Sabbatini observed, starting from the 1st century A.D., the settlement distribution is more homogeneously scattered across the entire survey sample area than it has ever been before (Marchi & Sabbatini 1996, 117 – 123; Sabbatini 2001, 72 – 75; Marchi 2004, 139; Marchi 2010, 43). The colonial center seems to be less important for the site locations; Imperial settlements exhibit a preference for the most favorable natural conditions for settlement and agriculture (*e.g.* fertile soils, ridges and peaks).

To conclude, in this paper we have cast new light on the evolution of the rural territory around the Roman town of Venusia. The main point we reached through our analyses is that settlement of the colonial period was less invasive and grew much more organically with respect to the pre-existing settlement patterns than is usually assumed. Indeed, the Roman colonization of the area in the 3rd century B.C. does not seem to have entailed a drastic reshaping of the previous territorial organization. Settlement location in the colonial period was directed more by the pre-existing sociopolitical organization of this territory, than by its environmental conditions (such as favorable soils). Regardless of whether the new sites of the colonial period can actually be linked directly to Roman settlers, or to indigenous or mixed strategies, this conclusion strongly questions traditional theories about Roman colonial settlement organization. In the past, scholars have lent great importance to natural constraints and in particular favorable soils for certain types of agriculture, and have disregarded the influence of native communities and their territorial organization as being irrelevant or indeed swept away by the Roman conquest (Salmon 1969; White 1970; Brown 1980; Rathbone 1981; Celuzza & Regoli 1982; Carandini *et al.* 2002, 108 – 110; see the critiques against this view by Terrenato 2001; 2007; Bradley 2006). Only in the 1st century B.C., and above all in the 1st century A.D. we see remarkable divergences from previous settlement patterns. Only at that point, natural conditions started playing a more important role in settlement strategies.

6.6 REFERENCES

Attema, P.A.J., Burgers, G.-J. & van Leusen, P.M. 2010. *Regional pathways to complexity: Settlement and land-use dynamics in early Italy from the Bronze age to the Republican period.* Amsterdam: Amsterdam University Press

Attwell, M. & Fletcher, M. 1985. A new technique for investigating spatial relationships: Significance testing. In A. Voorrips & S.H. Loving (eds.), *To pattern the past: Symposium on mathematical methods in archaeology, Amsterdam 1984*, 181 – 189. Strasbourg: Conseil de l'Europe.

Attwell, M.R. & Fletcher, M. 1987. An analytical technique for investigating spatial relationships. *Journal of Archaeological Science* 14(1), 1-11. doi: 10.1016/S0305-4403(87)80002-X

Azzena, G. & Tascio M. 1996. Il Sistema Informativo Territoriale per la carta archeologica d'Italia. In M.L. Marchi & G. Sabbatini, *Venusia (IGM 187 I NO / I NE)* (Forma Italiae 37), 281 – 297. Firenze: Leo S. Olschki

Barker, G. & Lloyd, J. (eds.) 1991. *Roman landscapes: archaeological survey in the Mediterranean region*. London: British School at Rome.

Barker, G. & Mattingly D.J. 1999. *The archaeology of Mediterranean landscapes*. Oxford: Oxbow Books.

Bintliff, J.L. & Sbonias, K.(eds.) 1999. *Reconstructing past population trends in Mediterranean Europe (3000 B.C.-A.D. 1800)*. Oxford: Oxbow Books.

Bintliff, J.L. & Sbonias, K. 2000. Demographic and ceramic analysis in regional survey. In: R. Francovich, H. Patterson & G. Barker (eds.), *Extracting meaning from ploughsoil assemblages*, 244-258. Oxford: Oxbow Books.

Bispham, E. 2006. Coloniam deducere: How Roman was Roman colonization during the Middle Republic? In G. Bradley & J.P. Wilson (eds.), *Greek and Roman colonization: Origins, ideologies and interactions*, 73-160. Swansea: Classical Press of Wales.

Bottini, A. 1980. L'area melfese dall'età arcaica alla romanizzazione (VI-III sec. a.C.). In *Attività archeologica in Basilicata 1964-1977. Studi in onore di Dinu Adamesteanu*, 313-344. Matera.

Bottini, A. 1982. Il melfese fra VII e V secolo a.C. *Dialoghi di Archeologia* 4(2), 152-160.

Bottini, A. 2001. Ripacandida. *BTCGI* XVI, 180-182.

Bottini, A. 2016. Popoli anellenici in Basilicata, mezzo secolo dopo. In M.L. Marchi (ed.), *Identità e conflitti tra Daunia e Lucania Preromane*, 7-50. Pisa: Edizioni ETS.

Bradley, G. 2006. Colonization and identity in Republican Italy. In G. Bradley & J.P. Wilson (eds.), *Greek and Roman Colonization: Origins, ideologies and interactions*, 161-187. Swansea: Classical Press of Wales.

Brown, F.E. 1980. *Cosa: The making of a Roman town*. Ann Arbor: University of Michigan Press.

Buck, R.J. 1971. The Via Herculia. *Papers of the British School at Rome* 39, 66 - 87.

Buck, R.J. 1981. The ancient roads of northwestern Lucania and the battle of Numistro. *La Parola del Passato* 36, 317-347.

Capogrossi Colognesi, L. 2002. *Persistenza e innovazione nelle strutture territoriali dell'Italia Romana. L'ambiguità di una interpretazione storiografica e dei suoi modelli*. Napoli: Jovene.

Carandini, A. 1994. I paesaggi agrari dell'Italia romana visti a partire dall'Etruria. In: *L'Italie d'Auguste à Dioclétien, Actes du colloque international de Rome (25-28 mars 1992)*

198 (1), 167-174. Rome: École Française de Rome.

Carandini, A., Cambi, F., Celuzza, M. G. & Fentress, E. 2002. *Paesaggi d'Etruria. Valle dell'Albegna, Valle d'Oro, Valle del Chiarone, Valle del Tafone*. Roma: Edizioni di Storia e Letteratura.

Casarotto, A. 2017. Location preferences of rural settlements in the territory of Venusia: an inductive approach. In H. Kamermans & C.C. Bakels (eds.), *Excerpta Archaeologica Leidensia II*, 163-209. Analecta Praehistorica Leidensia 47.

Casarotto, A., Pelgrom, J. & Stek, T.D. 2016. Testing settlement models in the early Roman colonial landscapes of Venusia (291 B.C.), Cosa (273 B.C.) and Aesernia (263 B.C.). *Journal of Field Archaeology* 41 (5), 568-586. doi:10.1080/00934690.2016.1211474

Casarotto, A., Stek, T.D., Pelgrom, J., van Otterloo, R.H., Sevink, J. 2018. Assessing visibility and geomorphological biases in regional field surveys: The case of Roman Aesernia. *Geoarchaeology, an International Journal* 33 (2): 177 – 192. doi: 10.1002/gea.21627

Celuzza, M. G. & Regoli, E. 1982. La valle d'Oro nel territorio di Cosa. Ager Cosanus e Ager Veientanus a confronto. *Dialoghi di Archeologia* 4, 31–62.

Conolly, J. & Lake, M. 2006. *Geographical Information Systems in archaeology*. Cambridge: Cambridge University Press.

Crawford, M.H. 1995. La storia della colonizzazione romana secondo i Romani. In A. Storchi Marino (ed.), *L'incidenza dell'antico. Studi in memoria di Ettore Lepore*. I, 187-192. Napoli: Luciano Editore

De Cazanove, O. 2005. Les colonies latines et les frontières régionales de l'Italie. *Venusia* et Horace entre Apulie et Lucanie: *Satires*, II, 1, 34. *Mélanges de la Casa de Velázquez* 35 (2), 107 – 124.

De Guio, A. 1985. Archaeological applications of survival analysis. In A. Voorrips & S.H. Loving (eds.), *To pattern the past: Symposium on mathematical methods in archaeology, Amsterdam 1984*, 361-381. Strasbourg: Conseil de l'Europe.

De Guio, A. 1988-1989. Analisi funzionale dei 'paesaggi di potere'. *Origini* XIV, 447 - 478.

De Guio, A., Evans, S.P. & Ruta Serafini, A. 1986. Marginalità territoriale ed evoluzione del "paesaggio di potere": un caso di studio nel Veneto. *Quaderni di Archeologia del Veneto* II, 160-172.

De Guio, A. & Secco, G. 1988. Archaeological applications of the 'Percolation Method' for data analysis and pattern recognition. In S.P.Q. Rahtz (ed.), *Computer and Quantitative Methods in Archaeology 1988, CAA88* (BAR International Series 446 (i)), 63 – 93. Oxford: BAR.

Dores Cruz, M. 2011. "Pots are pots, not people:" material culture and ethnic identity in the Banda Area (Ghana), nineteenth and twentieth centuries. *Azania: Archaeological Research in Africa* 46 (3), 336-357. doi: 10.1080/0067270X.2011.629525

Ducke, B. & Kroefges, P.C. 2008. From points to areas: constructing territories from archaeological site patterns using an enhanced Xtent model. In: A. Posluschny, K. Lambers & I. Herzog (eds.), *Layers of Perception. Proceedings of the 35th International Conference on Computer Applications and Quantitative Methods in Archaeology (CAA), Berlin, Germany, April 2–6, 2007* (Kolloquien zur Vor- und Frühgeschichte, Vol. 10), 245-251. Bonn 2008.

Eastman, J. R. 2012. *IDRISI Selva manual (Manual Version 17.01)*. Worcester, MA: Clark Labs, Clark University.

Eastman, J.R. 2016. *TerrSet manual*. Worcester, MA: Clark Labs, Clark University.

Esri, 2014. How Aspect works. In: *ArcGIS 10.2.2 Help*.

FAO, 2006. *Guidelines for Soil Description* (4th edition). Rome: Food and Agriculture Organization of the United Nations.

FAO, 2014. *World reference base for soil resources 2014. International soil classification system for naming soils and creating legends for soil maps* (Update 2015). Rome: Food and Agriculture Organization of the United Nations.

Fentress, E. 2000. What are we counting for? In

R. Francovich, H. Patterson & G. Barker (eds.), *Extracting meaning from ploughsoil assemblages*, 45 – 52. Oxford: Oxbow Books.

Fracchia, H. 2013. Survey, Settlement and Land Use in Republican Italy. In: J. DeRose Evans (ed.), *A Companion to the Archaeology of the Roman Republic*, 181-197. Blackwell Publishing Ltd.

Francovich, R., Patterson, H. & Barker, G. (eds.) 2000. *Extracting meaning from ploughsoil assemblages*. Oxford: Oxbow Books.

Frayn, J.M. 1979. *Subsistence Farming in Roman Italy*. Fontwell, Sussex: Centaur Press Ltd.

Gabba, E. 1977. Considerazioni sulla decadenza della piccola proprietà contadina nell'Italia centro-meridionale del II secolo a.C. *Ktema II*, 269-284.

Goodchild, H. 2007. *Modelling Roman agricultural production in the middle Tiber valley, Central Italy*. PhD dissertation, University of Birmingham: eTheses Repository. http://etheses.bham.ac.uk/175/

Goodchild, H. 2013. GIS models of Roman agricultural production. In A. Bowman & A. Wilson (eds.), *The Roman agricultural economy: organization, investment, and production*, 55-83. Oxford: Oxford University Press.

Gualtieri, M. 2002. Insediamenti e proprietà nella Lucania nord-orientale (I sec. a.C. – III sec. d.C.). In E. Lo Cascio & A. Storchi Marino (eds.), *Modalità insediative e strutture agrarie nell'Italia Meridionale in età Romana*, 75 – 102. Bari: Edipuglia.

Isayev, E. 2007. *Inside ancient Lucania. Dialogues in history and archaeology*. London: Institute of Classical Studies.

Kamermans, H. 2000. Land evaluation as predictive modelling: a deductive approach. In G. Lock (ed.), *Beyond the map: archaeology and spatial technologies*, 124-146. Amsterdam: IOS Press.

Kamermans, H. & Wansleeben, M. 1999. Predictive modelling in Dutch archaeology, joining forces. In J.A. Barceló, I. Briz & A. Vila (eds.), *New techniques for old times. CAA98. Computer applications and quantitative methods in archaeology. Proceedings of the 26th Conference, Barcelona, March 1998* (BAR International Series 757), 225-230. Oxford: Archaeopress.

Kvamme, K.L. 1990. One-sample tests in regional archaeological analysis: new possibilities through computer technology. *American Antiquity* 55(2), 367-381. doi: 10.2307/281655

Launaro, A. 2011. *Peasants and slaves. The rural population of Roman Italy (200 B.C. to A.D. 100)*. Cambridge: Cambridge University Press.

Marchi, M.L. 1991. Venosa: ceramica di tradizione non romana. In M. Salvatore (ed.), *Il Museo Archeologico Nazionale di Venosa*, 86-91. Matera: IEM editrice.

Marchi, M.L. 2000. Effetti del processo di romanizzazione nelle aree interne centro-meridionali. Acquisizioni, innovazioni ed echi tradizionali documentati archeologicamente. *Orizzonti* I, 227 - 242.

Marchi, M.L. 2004. Fondi, latifondi e proprietà Imperiali nell'Ager Venusinus. *Agri Centuriati* 1, 129 – 156.

Marchi, M.L. 2009. Modi e forme dell'urbanizzazione della Daunia. In M. Osanna (ed.), *Verso la città. Forme insediative in Lucania e nel mondo italico fra IV e III sec. a.C.*, 327-367. Venosa: Osanna Ed.

Marchi, M.L. 2010. *Ager Venusinus II (IGM 175 II SO; 187 I NO; 187 I SE; 188 IV NO; 188 IV SO)* (Forma Italiae 43). Firenze: Leo S. Olschki.

Marchi, M.L. 2014. The landscape of Daunia: Ager Venusinus. In A.M. Small (ed.), *Beyond Vagnari, New themes in the study of Roman South Italy*, 181-192. Bari: Edipuglia.

Marchi, M.L. 2016. Sanniti in Daunia. Dinamiche insediative tra VI e III secolo a.C. In M.L. Marchi (ed.), *Identità e conflitti tra Daunia e Lucania Preromane*, 51-84. Pisa: Edizioni ETS.

Marchi, M.L. & Sabbatini, G. 1996. *Venusia (IGM 187 I NO / I NE)* (Forma Italiae 37). Firenze: Leo S. Olschki.

Marchi, M.L. & Salvatore, M. 1997. *Venosa: forma e urbanistica* (Città antiche in Italia 5). Roma: «L'Erma» di Bretschneider.

Musti, D. 2009. *Lucanus an Apulus anceps*. Il territorio dei Lucani e i suoi confini fra il IV e il I sec. a.C. In M. Osanna (ed.), *Verso la città. Forme insediative in Lucania e nel mondo italico fra IV e III sec. a.C.*, 13 – 24. Venosa: Osanna Ed.

Nuninger, L., Verhagen, P., Bertoncello, F. & Castrorao Barba, A. 2016. Estimating "Land Use Heritage" to model changes in archaeological settlement patterns. *LAC 2014 Proceedings, 0*, 12. doi: http://dx.doi.org/10.5463/lac.2014.60

Odum, E.P. 1959. *Fundamentals of Ecology*. Philadelphia and London: W. B. Saunders Company.

Osanna, M. 2010. Paesaggi agrari e organizzazione del territorio in Lucania tra IV e III sec. a.C. *Bollettino di Archeologia on line, Volume Speciale*, 17-31.

Patterson, J. 2006. Colonization and historiography: the Roman Republic. In G. Bradley & J.P. Wilson (eds.), *Greek and Roman Colonization: Origins, ideologies and interactions*, 189 – 219. Swansea: Classical Press of Wales.

Pelgrom, J., Lecce, L., García Sánchez, J. & Stek, T.D. 2016. Le indagini olandesi sul pianoro di Masseria Casalini Sottana (2013 - 2014). In A. De Siena & T. Giammatteo (eds.), *Palazzo San Gervasio. Modalità insediative e pratiche funerarie dal territorio*, 277 – 288. Venosa: Osanna Edizioni.

Pelgrom, J., Marchi, M.L., Cantoro, G., Casarotto, A., Hamel, A., Lecce, L., García Sánchez, J. & Stek, T.D. 2014. New approaches to the study of village sites in the territory of Venosa in the Classical and Hellenistic periods. *Agri Centuriati* 11, 31-60. doi: 10.1400/233577

Pelgrom, J. & Stek, T.D. 2014. Roman Colonization under the Republic: historiographical contextualisation of a paradigm. In T.D. Stek & J. Pelgrom (eds.), *Roman Republican Colonization. New Perspectives from Archaeology and Ancient history*, 10-45. Rome: Palombi Editori.

Pelgrom, J. & Tetteroo, C. 2015. *Preliminary results from the Ager Venusinus project. Fieldreport of the 2013 and 2014 campaigns*. Internal Report for the Soprintendenza Archeologica della Basilicata.

Rathbone, D.W. 1981. The development of agriculture in the 'Ager Cosanus' during the Roman Republic: Problems of evidence and interpretation. *The Journal of Roman Studies* 71, 10-23. doi: 10.2307/299493

Rathbone, D.W. 2008. Poor peasants and silent sherds. In L. De Ligt & S.J. Northwood (eds.), *People, land and politics. Demographic developments and the transformation of Roman Italy, 300 B.C.-A.D. 14*, 305-332. Leiden: Brill.

Renfrew, C. & Level, E.V. 1979. Exploring dominance: Predicting polities from centers. In C. Renfrew & K.L. Cooke (eds.), *Transformations: Mathematical approaches to culture change*, 145-167. New York: Academic Press.

Roppa, A. & Van Dommelen, P. 2012. Rural settlement and land-use in Punic and Roman Republican Sardinia. *Journal of Roman Archaeology* 25, 49 – 68.

Sabbatini, G. 2001. *Ager Venusinus I. Mezzana del Cantore (IGM 175 II SE)* (Forma Italiae 40). Firenze: Leo S. Olschki.

Salmon, E.T. 1967. *Samnium and the Samnites*. Cambridge: Cambridge University Press.

Salmon, E.T. 1969. *Roman Colonization under the Republic*. London: Thames and Hudson.

Salvatore, M. (ed.) 1984. *Venosa: un parco archeologico ed un museo. Come e perchè*. Taranto: Editrice Scorpione.

Settis, S. (ed.) 1984. *Misurare la terra: centuriazione e coloni nel mondo romano*. Modena: Edizioni Panini.

Sewell, J. 2016. Higher-order settlements in early Hellenistic Italy: A quantitative analysis of a new archaeological database. *American Journal of Archaeology* 120 (4), 603 – 630. doi: 10.3764/aja.120.4.0603

Shennan, S. 1988. *Quantifying Archaeology*. Edinburgh: Edinburgh University Press.

Shennan, S.J. (ed.) 1994. Archaeological approaches to cultural identity. London: Routledge.

Siegel, S. 1956. *Nonparametric statistics for the behavioral sciences*. New York: McGraw-Hill.

Spurr, M.S. 1986. *Arable cultivation in Roman Italy c. 200 B.C. - c. A.D. 100* (Journal of Roman Studies Monographs, 3). London: Society for the Promotion of Roman Studies.

Stek, T.D. 2012. Review of: Marchi M. L. (2010) Ager Venusinus II. Forma Italiae 43. *BaBesch* 87, 243-244.

Stek, T.D. 2013. Material culture, Italic identities and the Romanization of Italy. In J. DeRose Evans (ed.), *A Companion to the Archaeology of the Roman Republic*, 337 – 353. Blackwell Publishing Ltd.

Stek, T. D. 2018. Early Roman colonization beyond the Romanizing agro-town: village patterns of settlement and highland exploitation in Central Italy. In B. S. Düring & T. D. Stek (eds.), *The Archaeology of Imperial Landscapes: A Comparative Study of Empires in the Ancient Near East and Mediterranean World*, 145-172. Cambridge: Cambridge University Press.

Stek, T. D., Modrall, E. B., Kalkers, R. A. A., van Otterloo, R. H. & Sevink, J. 2015. An early Roman colonial landscape in the Apennine mountains: Landscape archaeological research in the territory of Aesernia (Central-Southern Italy). *Analysis Archaeologica: An International Journal of Western Mediterranean Archaeology* 1, 229–291.

Stek, T.D. & Pelgrom, J. 2013. Landscapes of early Roman colonization: Non-urban settlement organization and Roman expansion in the Roman Republic (4th-1st centuries B.C.). *Tijdschrift voor Mediterrane Archeologie* 50, 87.

Tagliamonte, G. 1996. *I Sanniti: Caudini, Irpini, Pentri, Carricini, Frentani*. Milano: Longanesi.

Tarquini, S., Isola, I., Favalli, M., Mazzarini, F., Bisson, M., Pareschi, M. T. & Boschi, E. 2007. TINITALY/01: a new Triangular Irregular Network of Italy. *Annals of Geophysics* 50 (3), 407 - 425. doi: 10.4401/ag-4424

Tarquini, S. & Nannipieri, L. 2017. The 10 m-resolution TINITALY DEM as a trans-disciplinary basis for the analysis of the Italian territory: Current trends and new perspectives. *Geomorphology* 281, 108-115. doi: 10.1016/j.geomorph.2016.12.022

Tarquini, S., Vinci, S., Favalli, M., Doumaz, F., Fornaciai, A. & Nannipieri, L. 2012. Release of a 10-m-resolution DEM for the Italian territory: Comparison with global-coverage DEMs and anaglyph-mode exploration via the web. *Computers & Geosciences* 38 (1), 168-170. doi: 10.1016/j.cageo.2011.04.018

Terrenato, N. 2001. Introduction. In: S. Keay & N. Terrenato (eds.), *Italy and the West. Comparative issues in Romanization*, 1-6. Oxford: Oxbow Books.

Terrenato, N. 2007. The essential countryside: The Roman world. In S.E. Alcock & R. Osborne (eds.), *Classical archaeology*, 139-161. Malden, MA: Blackwell.

Terrenato, N. 2012. The enigma of "Catonian" villas: the *De agri cultura* in the context of second-century B.C. Italian architecture. In: J.A. Becker & Terrenato N. (eds.), *Roman Republican villas: architecture, context and ideology*, 69-93. Ann Arbor: The University of Michigan Press.

Torelli, M. 1990. La formazione della villa. In G. Clemente, F. Coarelli & E. Gabba (eds.), *Storia di Roma. 2. L'impero mediterraneo. I. La Repubblica Imperiale*, 123-132. Torino: Einaudi.

Torelli, M. 1991. La fondazione di Venosa nel quadro della romanizzazione dell'Italia meridionale. In: M. Salvatore (ed.), *Il Museo archeologico nazionale di Venosa*, 17–26. Matera: IEM editrice.

Torelli, M. 1992. Venosa romana. In P. Fedeli (ed.), *Venosa*, 35 – 77. Venosa

Torelli, M. 1999. *Tota Italia: Essays in the cultural formation of Roman Italy*. Oxford: Clarendon Press.

Van Dommelen, P. & Terrenato, N. 2007. Local cultures and the expanding Roman Republic. In P. Van Dommelen & N. Terrenato (eds.), *Articulating local cultures, power and identity under the expanding Roman Republic*, 7 – 12. Journal of Roman archaeology, JRA supplementary series number 63. Portsmouth.

Van Joolen, E. 2003. *Archaeological land evaluation.*

A reconstruction of the suitability of ancient landscapes for various land uses in Italy focused on the first millennium B.C. PhD dissertation. University of Groningen: University of Groningen Research Database. http://hdl.handle.net/11370/78b6a5bf-917d-4a15-969d-230c3c72b690

Van Leusen, P.M. 2002. *Pattern to process: Methodological investigations into the formation and interpretation of spatial patterns in archaeological landscapes.* PhD dissertation. University of Groningen: University of Groningen Research Database http://hdl.handle.net/11370/fc9ead23-e2c7-40c5-942f-e06f52198222

Van Leusen, P.M. & Kamermans, H. (eds.) 2005. *Predictive modelling for archaeological heritage management: A research agenda.* Nederlandse Archeologische Rapporten, vol. 29. Amersfoort: Rijksdienst voor het Oudheidkundig Bodemonderzoek.

Vermeulen, F 2017. *From the mountains to the sea. The Roman colonisation and urbanisation of Central Adriatic Italy*. Babesch Supplements 30. Leuven: Peeters.

Vink, A.P.A. 1975. *Land use in advancing agriculture.* Berlin Heidelberg New York: Springer – Verlag.

Vinson, P. 1979. Il percorso della Via Appia tra Venosa e Palazzo S. Gervasio. *Lucania Archeologica* 3, 15-18.

Vita-Finzi, C. & Higgs, E.S. 1970. Prehistoric economy in the Mount Carmel area of Palestine: Site catchment analysis. *Proceedings of the Prehistoric Society* 36, 1–37.

Volpe, G. 1990. *La Daunia nell'età della romanizzazione: paesaggio agrario, produzione, scambi.* Bari: Edipuglia.

Wheatley, D. & Gillings, M. 2002. Spatial technology and archaeology. The archaeological applications of GIS. London & New York: Taylor and Francis.

White, K.D. 1970. *Roman farming*. London: Thames and Hudson.

Witcher, R. 2008. (Re)surveying Mediterranean rural landscapes: GIS and legacy survey data. *Internet Archaeology* 24. doi: 10.11141/ia.24.2

Wood, J. 2009. The LandSerf manual. http://www.staff.city.ac.uk/~jwo/landserf/landserf230/doc/landserfManual.pdf

Chapter 7. Conclusions

This book proposes a GIS-based procedure to use (legacy) survey data for regional-scale investigations of settlement patterns. Considering the wealth and growing accessibility of these data, it is vital to develop solid methodologies that permit us to use this complex and variegated dataset for historical reconstructions. To this effect, this study has presented a method which not only offers a way to formally tackle a research question using (legacy) site-based survey data (*i.e.* deductive-inductive strategy), but which also provides widely-applicable means to deal with biases affecting these data. The study of surface visibility and geomorphological processes needs to become standard procedure in regional settlement pattern analyses (see discussion and applications in *e.g.* Terrenato & Ammerman 1996; van Leusen 2002; Banning *et al.* 2011; 2017; Feiken 2014; De Neef 2016; Sevink *et al.* 2016). This book underlines the advantages that the assessment of biases can provide for the interpretation of settlement patterns. The analysis of distortions in site patterns allows researchers to recognize misleading observations and avoid biased interpretations. Clearly, this type of analysis should be incorporated as an essential component within the landscape archaeology project work plan (see discussion in van Leusen 2002; Terrenato 2004). In this way, a distinction can be made between those patterns produced by post-depositional processes and those that are instead representative of the ancient settlement organization. The method proposed in this book incorporates this bias-testing as a key step in the procedure of reconstructing regional settlement patterns from site-based survey data.

Regarding the case-study, we have seen that the method allowed answering the central research question, on which of the two proposed scenarios of early Roman colonial rural settlement organization is most likely to be correct. By applying a deductive-inductive procedure, new light could be shed on early Roman colonial landscapes in central and southern Italy. This research strategy not only permitted efficient testing of the reliability of competing settlement theories on an inter-regional level (deductive analysis, Chapter 3); it also offered a better understanding of local variations in settlement preferences that further enhanced our understanding of early colonial period settlement behavior in specific regions (inductive analysis, Chapter 6). Bias-testing (Chapters 4 and 5) allowed us to exclude the possibility that survey site patterns were the mere result of modern visibility and geomorphological factors, rather than representative of ancient settlement strategies.

Two significant conclusions on settlement patterns in early Roman colonial territories were reached. First, the conventional colonial settlement model which presumes that colonial rural settlements were regularly organized, has been seriously undermined. The detailed analysis presented in this book has shown that this theory is incompatible with the available survey site-based data which, for the treated case-study, consisted of recorded site point distributions (for other studies that reached similar conclusions for these but also for other contexts using different approaches and data see Pelgrom 2008; Stek 2009; 2018). In fact, this study provided solid evidence that the survey data instead strongly match the more recently proposed polynuclear, village-based settlement theory in terms of settlement density and site patterns. Based on the different interrelated analyses carried out in this book, it can now be excluded that the clustered patterns in the dataset of Aesernia and Venusia are the result of post-depositional biases or of major methodological differences in survey recording methods (for another colonial case-study with similar results see Hayes & Martini 1994, esp. ch. 3). The clusters of sites, as well as the low site density areas, and the empty zones in between them, are most likely the result of ancient location preferences and connected settlement strategies, rather than the result of the visibility and geomorphological biases considered in this analysis.

The second conclusion regards one of the contexts analyzed in this book, namely the territory of the Latin colony of Venusia. For this region, it was possible to dig deeper into the logic behind the recorded site patterns. Chapter 6 showed that the settlement clustering of archaeological sites did not happen randomly across the landscape, but instead targeted specific land pockets only marginally settled in the previous, pre-colonial period. According to the results from the inductive analysis, the preference for these zones seems to have been determined by two main cultural factors during the colonial period (3rd – 1st century B.C.): namely, a) the pre-colonial settlement organization and b) the vicinity to the colonial urban center. The influence of the latter factor was already noted during the deductive analysis performed earlier in Chapter 3 (section 3.4.1), when the data were compared with the 'adjusted' Von Thünen model.

The working hypothesis of a nucleated settlement organization could thus be confirmed (deductive analysis, Chapter 3) and complemented with new information regarding the rationale that probably determined the location of settlement clusters in the territory of Venusia (inductive analysis, Chapter 6). Of course, it is impossible to tell from patterns in survey sites whether the people living there were of Roman or native origin. However, the research presented in this book has demonstrated that during the Republican period the communities settling these territories adopted a settlement strategy that significantly differed from conventional expectations. These communities seem to have placed their settlements in a flexible way by choosing to cluster their dwellings in those zones only marginally settled in the previous pre-colonial period. This choice seems to be independent from favorable environmental factors, or at least these were not the most influential aspects. Instead, settlement location was likely related to the influence played by the colonial town and the geopolitical landscape organization (*i.e.* local cultural constraints).

These newly acquired insights on the character of rural settlement patterns in early colonial landscapes of central and southern Italy can now be tested further in the field or in the laboratory, as well as being applicable to testing for survey datasets of other territories. This study has demonstrated that, if approached in a proper way, (legacy) survey data can disclose important information about settlement patterns and about the principal processes that determined these patterns.

7.1 REFERENCES

Banning, E.B., Hawkins, A. L. & Stewart, S. T. 2011. Sweep widths and the detection of artifacts in archaeological survey. *Journal of Archaeological Science* 38 (12), 3447-3458. doi: 10.1016/j.jas.2011.08.007

Banning, E.B., Hawkins, A.L., Stewart, S.T., Hitchings, P. & Edwards, S. 2017. Quality Assurance in Archaeological Survey. *Journal of Archaeological Method and Theory* 24(2), 466 – 488. doi: 10.1007/s10816-016-9274-2

De Neef, W. 2016. *Surface <> Subsurface: A methodological study of Metal Age settlement and land use in Calabria (Italy).* PhD dissertation, University of Groningen: University of Groningen Research Database. http://hdl.handle.net/11370/d2b83a4f-3afd-4640-a614-c3bfb2bc5f76

Feiken, M. 2014. *Dealing with biases: Three Geo-Archaeological approaches to the hidden landscapes of Italy*. Eelde, Netherlands: Barkhuis.

Hayes, J. W. & Martini, I. P. 1994. *Archaeological survey in the lower Liri Valley, Central Italy*. Oxford.

Pelgrom, J. 2008. Settlement organization and land distribution in Latin colonies before the Second Punic War. In L. De Ligt &

S. J. Northwood (eds.), *People, land and politics. Demographic developments and the transformation of Roman Italy 300 B.C.-A.D. 14*, 317–356. Leiden: Brill.

Sevink, J., den Haan, M. & van Leusen M. 2016. *Soils and Soil Landscapes of the Raganello river catchment (Calabria, Italy)*. (Raganello Basin Studies 2). Eelde: Barkhuis Publishing & Groningen Institute of Archaeology.

Stek, T.D. 2009. *Cult places and cultural change in Republican Italy. A contextual approach to religious aspects of rural society after the Roman conquest*. Amsterdam: Amsterdam University Press.

Stek, T. D. 2018. Early Roman colonization beyond the Romanizing agro-town: village patterns of settlement and highland exploitation in Central Italy. In B. S. Düring & T. D. Stek (eds.), *The Archaeology of Imperial Landscapes: A Comparative Study of Empires in the Ancient Near East and Mediterranean World*, 145-172. Cambridge: Cambridge University Press.

Terrenato, N. 2004. Sample size matters! The paradox of global trends and local surveys. In S.E. Alcock & J.F. Cherry (eds.), *Side-by-side survey: Comparative regional studies in the Mediterranean world*, 36-48. Oxford: Oxbow Books.

Terrenato, N. & Ammerman, A. J. 1996. Visibility and site recovery in the Cecina valley survey, Italy. *Journal of Field Archaeology* 23 (1), 91-109. doi: 10.1179/009346996791973990

Van Leusen, P. M. 2002. *Pattern to process: Methodological investigations into the formation and interpretation of spatial patterns in archaeological landscapes*. PhD dissertation, University of Groningen: University of Groningen Research Database. http://hdl.handle.net/11370/fc9ead23-e2c7-40c5-942f-e06f52198222

Appendix A

Table A.1 Legend of the soil map (Fig. 4.3, p. 79) with the indication of the Hellenistic detectability for each soil unit (Fig. 4.4, p. 80). The soil description of units 1 to 41 is based on Koopmans (1980) and the description of units 43 to 66 is based on van Otterloo (1981; Stek et al. 2015, 291). The references for Koopmans 1980; van Otterloo 1981; Stek *et al.* 2015 are in section 4.8 of Chapter 4. Notes: the asterisk beside the soil name indicates that volcanic material was recorded or is likely present. For soil unit 6, archaeological material might be present in the colluvium coming from the surrounding mountains, thereby the detectability is considered 'medium'. This table is also provided as supporting information in the paper Casarotto, A., Stek, T.D., Pelgrom, J., van Otterloo, R.H. & Sevink, J. 2018. Assessing visibility and geomorphological biases in regional field surveys: The case of Roman Aesernia. *Geoarchaeology, an International Journal* 33 (2), 177 – 192. https://doi.org/10.1002/gea.21627

Landscape and geological system	Physiography	Soil unit	Soil name	Topography	FAO/UNESCO classification (1974)	Late Holocene gradational processes and scores	Hellenistic detectability
Fluviatile	Older and recent fluviatile accumulation on terraces	1	A1a	Flat to gently sloping (terrace partly covered by colluvial deposits)	Complex Orthic Luvisol/ Calcic Cambisol	deposition: 2	Medium
		2	A1b	Flat to gently sloping (terrace)	Complex Chromic Luvisol/ Calcic Cambisol	deposition: 1	Medium/High
		3	A1c	Flat or almost flat (terrace partly covered by colluvial deposits, channel migrations occur)	Complex Calcic Cambisol/ Calcaric Fluvisol	deposition: 2 erosion: 2	Medium/Low
		4	A1d	Flat or almost flat (young alluvial deposits, channel migrations occur)	Calcaric Fluvisol	erosion: 2	Low
Karst basins	Recent large basin deposits	5	B1a	Flat to gently sloping (basin deposit)	Orthic Luvisol	deposition: 3	Low
	Recent smaller basin deposits	6	B2a	Flat to gently sloping (basin and colluvial deposits)	Complex Calcaro-Gleyic Phaeozem/ Calcic Cambisol	deposition: 3	Medium
		7	B2b	Flat or almost flat (basin deposit)	Calcaric Gleysol	deposition: 3	Low
		8	B2c	Flat or almost flat (basin deposit)	Complex Calcaric Gleysol/ Calcaric Fluvisol	deposition: 3	Low

Dolomitic	Colluvial deposits from the surronding mountains	9	D2	Gently sloping to sloping (colluvial-alluvial fans)	Complex Calcic Cambisol/ Calcaric Regosol	deposition: 2 erosion: 2	Medium
Fluvio-lacustrine	Sub-horizontal clastic fluvio-lacustrine or travertine deposits	10	F1a	Sub-horizontal	Eutric Nitosol	erosion: 1	High
		11	F1b	Flat to gently sloping (eroded phase of F1a)	Complex Orthic Luvisol/ Calcic Luvisol	erosion: 2	Medium
		12	F1c	Flat to gently sloping (more eroded phase of F1a)	Complex Calcic Luvisol/ Luvic Phaeozem	erosion: 3	Medium
		13	F1d	Flat to gently sloping (eroded)	Complex Calcic Luvisol/ Calcaro-Luvic Phaeozem	erosion: 2	Medium
		14	F1e	Flat to gently sloping	Complex Calcaro-Luvic Phaeozem/ Chromic Vertisol	deposition and erosion: 0	High
	Dissected fluvio-lacustrine or travertine deposits	15	F2a	Gently sloping to sloping (colluvial deposits)	Complex Calcaro-Luvic Phaeozem/ Calcic Cambisol	deposition: 2 erosion: 1	Medium/Low
		16	F2b	Gently sloping (colluvial deposits)	Calcic Cambisol	deposition: 2 erosion: 1	Medium/Low
		17	F2c	Gently sloping to sloping (colluvial deposits in gullies)	Complex Calcaro-Luvic Phaeozem/ Calcic Cambisol	deposition: 2 erosion: 1	Medium/Low
		18	F2d	Gently sloping to sloping (eroded fluvio-lacustrine deposits)	Complex Calcic Luvisol/ Calcic Cambisol	deposition: 2 erosion: 2	Medium
	Steep slopes of the dissected fluvio-lacustrine landscape	19	F3a	Moderately steep with downward colluvial deposits	Complex Calcic Cambisol/ Calcaric Phaeozem	deposition: 2 erosion: 3	Low
		20	F3b	Moderately steep with more downward colluvial deposits	Complex Calcaric Phaeozem/ Calcic Cambisol	deposition: 2 erosion: 3	Low
	Terraces of the fluvio-lacustrine deposits	21	F4	Flat to gently sloping	Orthic Luvisol	deposition and erosion: 0	High
Limestone	Slopes of the limestone mountains	22	K1a	Moderately steep to steep	Lithosol	deposition: 0/1 erosion: 3	Low
		23	K1e	Sloping to moderately steep	Complex Lithosol/ Rendzina	deposition: 0/1 erosion: 3	Low

		24	K1f*	Moderately steep (covered by forest)	Mollic Andosol	deposition: 0/1 erosion: 1	Medium/High
		25	K1g	Steep (dissected)	Calcaric Regosol	erosion: 2	Medium/Low
		26	K1h	Moderately steep to steep (dissected)	Complex Rendzina/ Calcaric Regosol	erosion: 3	Low
		27	K1j	Moderately steep	Complex Rendzina/ Lithosol	erosion: 2	Medium/Low
	Accumulation zones of the limestone mountains	28	K2a*	Sub-horizontal (karst depressions)	association Orthic Luvisol/ Calcic Cambisol	deposition: 1 erosion: 1	Medium/High
		29	K2c*	Gently sloping to sloping (alluvial fans)	Chromic Luvisol	erosion: 1	High
		30	K2e	Sloping (colluvial deposits)	association Chromic Luvisol/ Chromic-Calcic Cambisol	deposition: 1	Medium/High
		31	K2f	Moderately sloping (colluvial deposits)	Vertic-Calcic Cambisol	deposition: 2	Medium/Low
Marls and shales	Slopes where recent erosion is dominant	32	M1a	Steep	Complex Rendzina/ Calcaric Regosol	erosion: 3	Low
		33	M1b	Moderately steep to steep	association Calcaric Regosol/ Calcic Cambisol	erosion: 3	Low
		34	M1c	Sloping to moderately steep	association Calcaric Regosol/ Calcic Cambisol	erosion: 3	Low
	More stable slopes	35	M2a	Gently sloping to sloping	Complex Orthic Luvisol/ Calcic Cambisol	erosion: 2	Medium/Low
		36	M2b	Gently sloping to sloping	association Calcic Cambisol/ Calcaric Regosol	erosion: 2	Medium/Low
		37	M2c	Sloping to moderately steep	association Rendzina/ Calcaric Regosol/ Chromic-Calcic Cambisol	erosion: 2	Medium/Low
	Relatively stable surfaces	38	M3a	Gently sloping to sloping (slightly dissected alluvial fans)	Calcic Cambisol	deposition: 2 erosion: 1	Low
		39	M3b	Gently sloping to sloping (dissected colluvial deposits)	association Chromic Luvisol/ Calcic Cambisol	deposition: 1	Medium/High

		40	M3c	Gently sloping to sloping (gullies with colluvial deposits)	Calcic Cambisol	deposition: 3	Medium/Low
		41	M3d*	Flat to gently sloping (karst depressions filled with colluvial deposits)	Chromic Vertisol	deposition: 1	High
Urban area		42					Low
Fluviatile	Higher accumulation terrace consisting of recent and sub-recent fluviatile sediments	43	F1a*	Level to nearly level (accumulation terrace)	Complex Calcic Cambisol/ Chromic-Calcic Cambisol	deposition: 2	High
		44	F1b*	Level to nearly level (accumulation terrace)	Calcic Cambisol	deposition: 2	High
	Surfaces between the terrace and the riverbed	45	F2a	Level to nearly level (occasionally flooded)	Complex Calcaric Fluvisol/ Calcic Cambisol	deposition: 2	Low
		46	F2b	Level to nearly level (occasionally flooded)	Complex Calcaric Fluvisol/ Calcic Cambisol	deposition: 2	Low
	Riverbed	47	F3a	Level to nearly level (riverbed)	Calcaric Fluvisol	erosion: 3 deposition: 2	Low
		48	F3b	Level to nearly level (riverbed)	Calcaric Fluvisol	erosion: 3 deposition: 2	Low
Fluvio-lacustrine	Old and relatively stable surfaces	49	L1a*	Level to gently sloping (terrace, covered by forest)	Complex Calcic Luvisol/ Calcic-Gleyic Luvisol	deposition: 1 erosion: 1	High
		50	L1b*	Level to gently sloping (terrace)	Complex Vertic Gleyic Luvisol/ Vertic-Calcic Luvisol	deposition: 1 erosion: 1/2	Medium/High
		51	L1c*	Level to gently sloping (terrace, covered by forest)	Complex Calcaric-Luvic Phaeozem/ Chromic-Gleyic Luvisol	deposition: 1 erosion: 0/1	High
		52	L1e*	Gently sloping (terrace)	association Vertic Luvisol/ Orthic Luvisol	deposition: 0/1 erosion: 1/2	Medium/High
	Middle and lower erosion terraces	53	L2a*	Level to gently sloping (terrace with colluvial deposits)	association Vertic-Gleyic-Luvisol/ Chromic-Gleyic Luvisol	deposition: 1 erosion: 0/1	High
		54	L2b*	Level to gently sloping (eroded terrace)	association Vertic Luvisol/ Vertic-Gleyic-Luvisol	deposition: 1 erosion: 2	Medium
		55	L2c*	Level to gently sloping (terrace with colluvial deposits)	Vertic Luvisol	deposition: 2 erosion: 1	Medium

		56	L2d*	Level to gently sloping (eroded terrace)	Complex Orthic Luvisol/ Calcic Cambisol	deposition: 1 erosion: 2	Medium
	Slopes of the Volturno and Vandra rivers	57	L3a*	Moderately steep to steep	Complex Vertic Luvisol/ Vertic-Gleyic Luvisol	erosion: 2	Medium/Low
		58	L3b	Steep to very steep	association Calcaric Regosol/ Orthic Luvisol	erosion: 2/3	Medium/Low
		59	L3c	Steep to very steep	Complex Calcaric Regosol/ Calcic-Gleyic Cambisol	erosion: 3	Low
	Recent and sub-recent dissected fluvio-lacustrine sediments	60	L4a*	Moderately steep (alluvial-colluvial deposits)	Complex Vertic Luvisol/ Vertic-Gleyic Luvisol	erosion: 1 deposition: 0/1	Medium/High
		61	L4b*	Step to very steep (dissections)	association Calcaric Regosol/ Orthic Luvisol	erosion: 1/2	Medium
		62	L4c*	Level to gently sloping (dissections)	Orthic Luvisol	erosion: 1	Medium
Marls and shales	Relatively stable surfaces	63	M1a	Gently sloping	Calcic Luvisol	erosion: 1	High
	Instable and strongly dissected surfaces	64	M2a	Moderately steep to steep	Complex Calcaric Regosol/ Calcic Cambisol	erosion: 3	Low
	Very instable and strongly dissected surfaces	65	M4a	Very steep	Complex Lithosol/ Calcaric Regosol	erosion: 3	Low
	Relatively stable colluvial deposits	66	M7c*	Gently sloping to sloping (colluvial deposits)	association Vertic-Calcic Cambisol/ Calcaric Regosol	deposition: 2	Medium

Appendix B

Results of the inductive analysis

This appendix is composed by three parts. The first part (a) displays the graphs related to the two comparisons carried out in Chapter 6 (section 6.4). These graphs represent the settlement distributions per period for each variable considered in the inductive analysis. The second part (b) reports the graphs of the settlement distributions per period and per size for each variable considered in the inductive analysis. In the third part (c) the tables show the results of the statistical tests applied in the inductive analysis to explore possible correlations. These tools

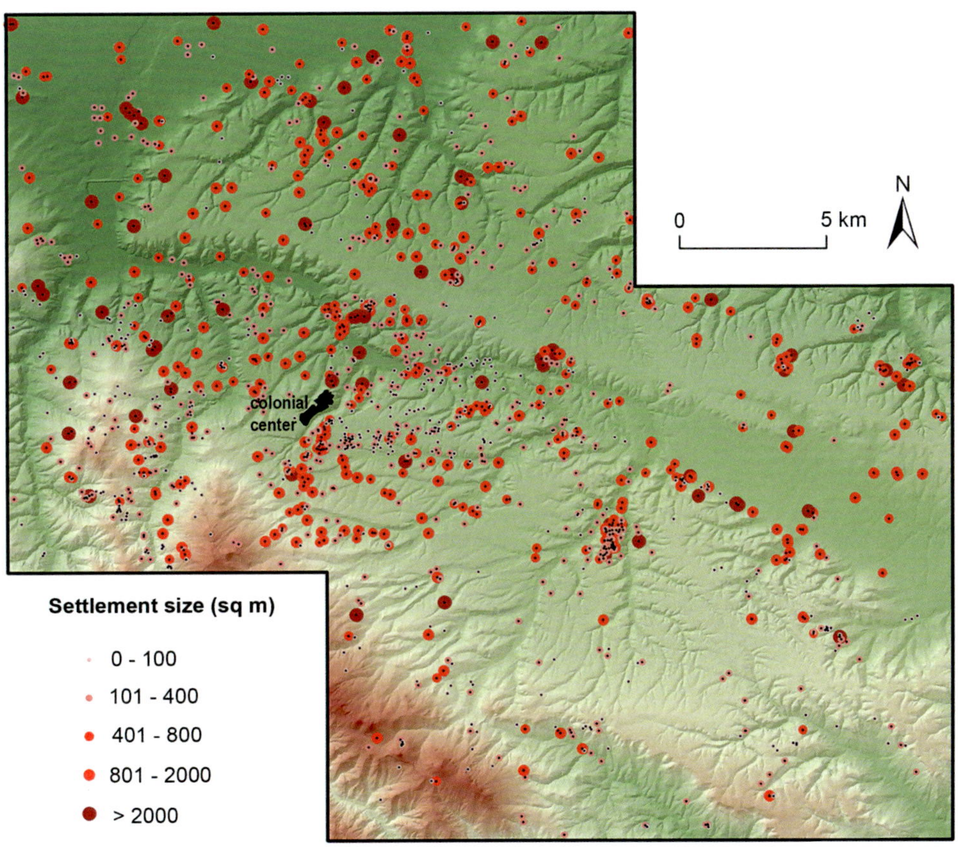

Fig. B.1 Total settlement distribution (pre-Roman to Imperial period settlements). The position of the settlements is indicated by black dots. The extension of the red circles circumscribing these black dots does not match with the scale of the map; they are used here only as symbols for the size of these settlements (see legend). The raster base map is the shaded relief calculated from the 10 m-resolution DEM named TINITALY/01 (Tarquini *et al.* 2007; 2012; Tarquini & Nannipieri 2017).

(*i.e.* graphs and statistical tests)[32] were used together in the inductive analysis to identify the significant correlations (described in Chapter 6 section 6.4).

32 Figures B.1, B.6-13 and Tables B.1 - 5 are also published in Casarotto, A. 2017. Location preferences of rural settlements in the territory of Venusia: an inductive approach. In: H. Kamermans & C. C. Bakels (eds.), *Excerpta Archaeologica Leidensia II*, 163-209. Analecta Praehistorica Leidensia 47.
Figures B.2-5 are also provided as Online Resource in the paper Casarotto, A., Pelgrom, J. & Stek, T.D. 2017. A systematic GIS-based analysis of settlement developments in the landscape of Venusia in the Hellenistic-Roman period. *Archaeological and Anthropological Sciences.* https://doi.org/10.1007/s12520-017-0561-y

a) *Distribution graphs per period*

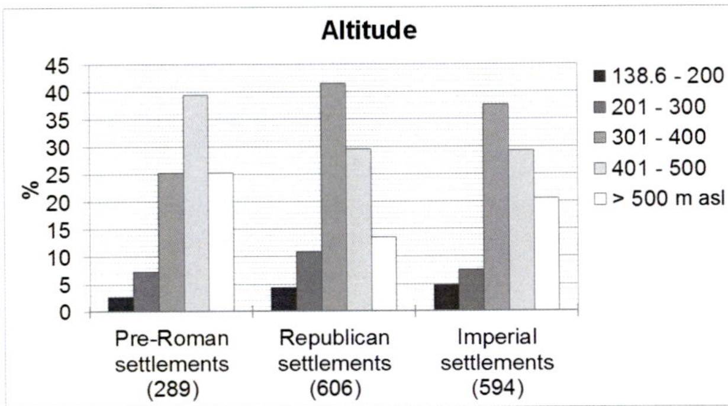

Altitude band		Area (sq m)	%
1st	138.6 - 200	42809500	6.2
2nd	201 - 300	78911100	11.3
3rd	301 - 400	236098000	34.0
4th	401 - 500	179970600	25.9
5th	> 500 m a.s.l.	157493000	22.7
Tot.		695282200	100

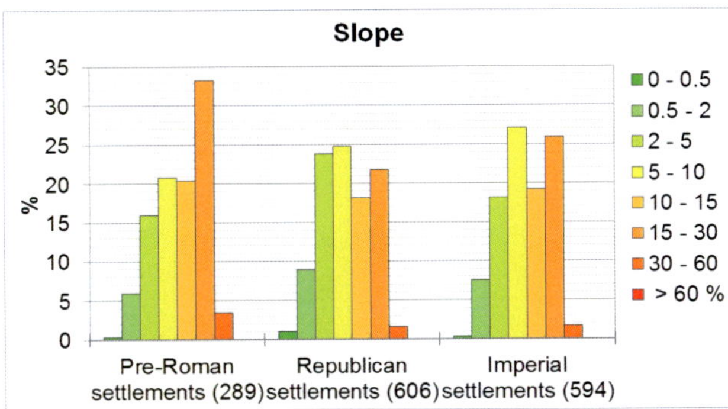

Slope class		Area (sq m)	%
1st	0 - 0.5	3679400	0.53
2nd	0.5 - 2	79234200	11.40
3rd	2 - 5	130012400	18.70
4th	5 - 10	121022700	17.41
5th	10 - 15	107931100	15.52
6th	15 - 30	201213300	28.94
7th	30 - 60	49488800	7.12
8th	> 60 %	2700300	0.39
Tot.		695282200	100

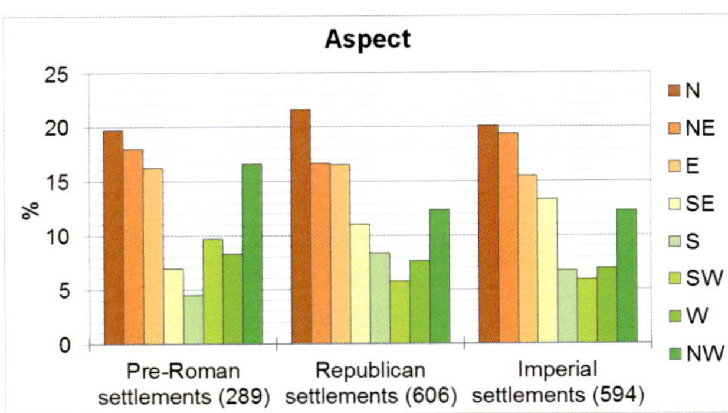

Aspect class		Area (sq m)	%
	Flat	63400	0.01
1st	N	143747900	20.67
2nd	NE	112804900	16.22
3rd	E	84013500	12.08
4th	SE	74520900	10.72
5th	S	60661100	8.72
6th	SW	55319500	7.96
7th	W	64981400	9.35
8th	NW	99169600	14.26
Tot.		695282200	100

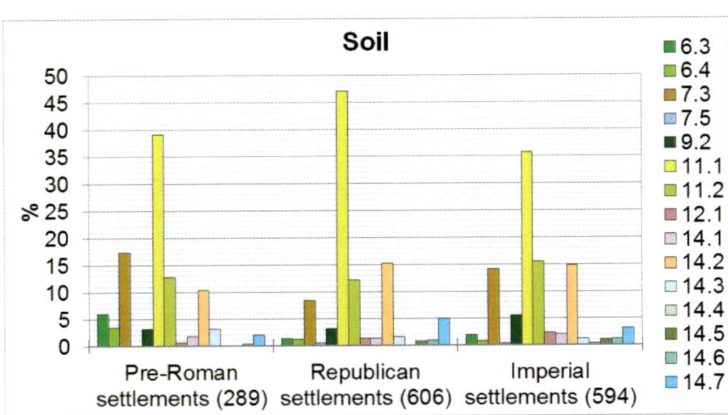

Soil unit	Area (sq m)	%
6.3	13156400	1.89
6.4	6690000	0.96
7.3	67255200	9.67
7.5	7361700	1.06
9.2	15808800	2.27
11.1	279720100	40.23
11.2	135817200	19.53
12.1	23297100	3.35

Soil unit	Area (sq m)	%
14.1	28993500	4.17
14.2	61494600	8.84
14.3	8919300	1.28
14.4	2542200	0.37
14.5	9745700	1.40
14.6	15408200	2.22
14.7	19072200	2.74
Tot.	695282200	100

Fig. B.2 Graphs of the percentages of pre-Roman, Republican, and Imperial settlements with respect to altitude, slope, aspect and soil factors (cf. section 6.4.1 of Chapter 6). In brackets, number of total settlements per period.

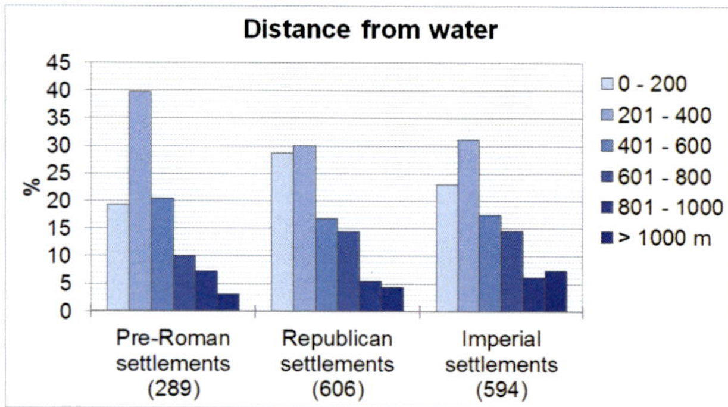

Water band		Area (sq m)	%
1st	0 - 200	175648200	25.26
2nd	201 - 400	164570200	23.67
3rd	401 - 600	127113900	18.28
4th	601 - 800	89448700	12.87
5th	801 - 1000	56637100	8.15
6th	> 1000 m	81864100	11.77
Tot.		695282200	100

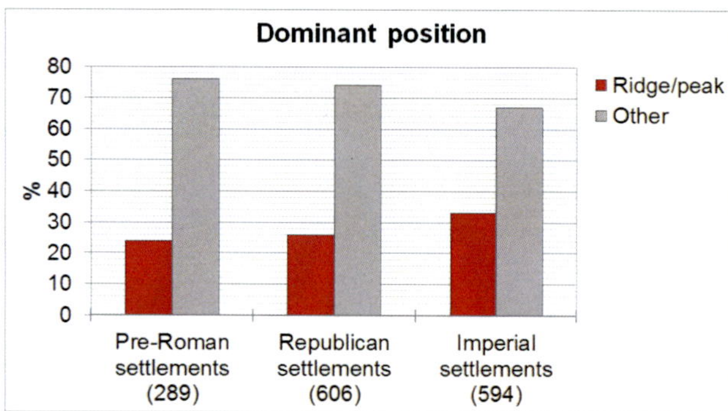

Dominant position	Area (sq m)	%
Ridge/Peak	129836500	18.67
Other	565445700	81.33
Tot.	695282200	100

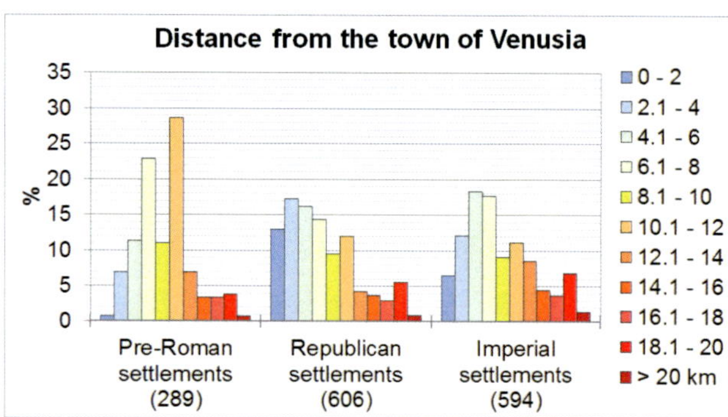

Distance band		Area (sq m)	%
1st	0 - 2	20050700	2.88
2nd	2.1 - 4	44602500	6.42
3rd	4.1 - 6	66629100	9.58
4th	6.1 - 8	82483500	11.86
5th	8.1 - 10	99370200	14.29
6th	10.1 - 12	97895100	14.08
7th	12.1 - 14	88587200	12.74
8th	14.1 - 16	60777500	8.74
9th	16.1 - 18	43986300	6.33
10th	18.1 - 20	41855200	6.02
11th	> 20 km	49044900	7.05
Tot.		695282200	100

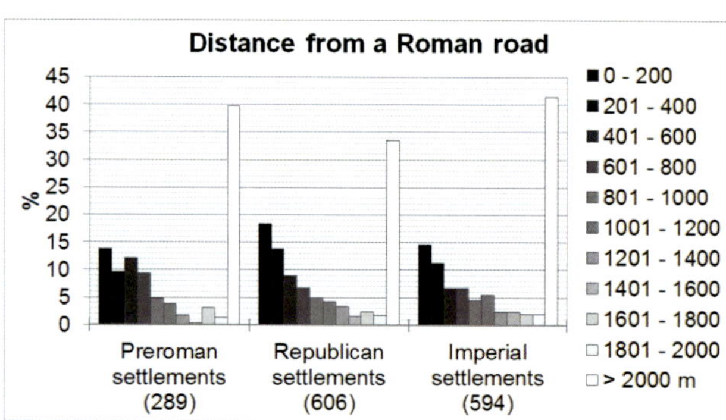

Distance band		Area (sq m)	%
1st	0 - 200	51136700	7.35
2nd	201 - 400	46892800	6.74
3rd	401 - 600	40638300	5.84
4th	601 - 800	37719400	5.43
5th	801 - 1000	34711500	4.99
6th	1001 - 1200	31034200	4.46
7th	1201 - 1400	27170200	3.91
8th	1401 - 1600	23640300	3.40
9th	1601 - 1800	21404100	3.08
10th	1801 - 2000	20094000	2.89
11th	> 2000 m	360840700	51.90
Tot.		695282200	100

Fig. B.3 Graphs of the percentages of pre-Roman, Republican, and Imperial settlements with respect to water distance, the location of dominant positions, the distance from the town of Venusia and Roman roads (cf. section 6.4.1 of Chapter 6). In brackets, number of total settlements per period.

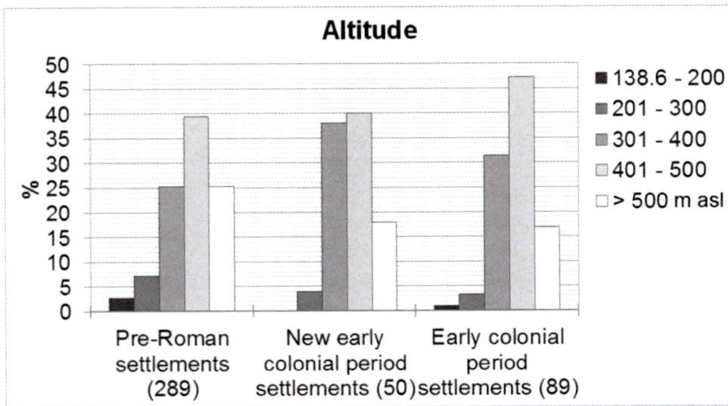

Altitude band		Area (sq m)	%
1st	138.6 - 200	42809500	6.2
2nd	201 - 300	78911100	11.3
3rd	301 - 400	236098000	34.0
4th	401 - 500	179970600	25.9
5th	> 500 m a.s.l.	157493000	22.7
Tot.		695282200	100

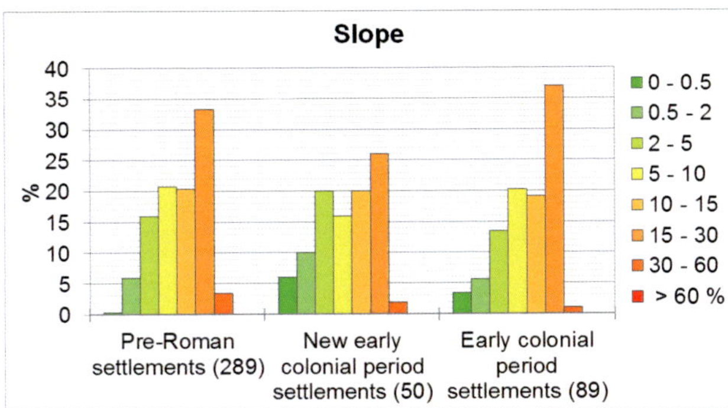

Slope class		Area (sq m)	%
1st	0 - 0.5	3679400	0.53
2nd	0.5 - 2	79234200	11.40
3rd	2 - 5	130012400	18.70
4th	5 - 10	121022700	17.41
5th	10 - 15	107931100	15.52
6th	15 - 30	201213300	28.94
7th	30 - 60	49488800	7.12
8th	> 60 %	2700300	0.39
Tot.		695282200	100

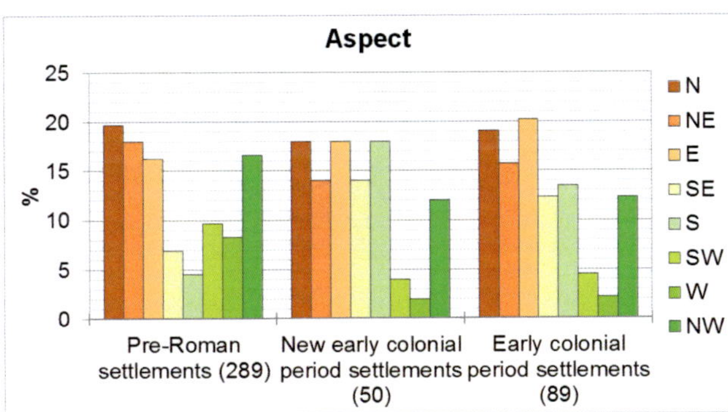

Aspect class		Area (sq m)	%
	Flat	63400	0.01
1st	N	143747900	20.67
2nd	NE	112804900	16.22
3rd	E	84013500	12.08
4th	SE	74520900	10.72
5th	S	60661100	8.72
6th	SW	55319500	7.96
7th	W	64981400	9.35
8th	NW	99169600	14.26
Tot.		695282200	100

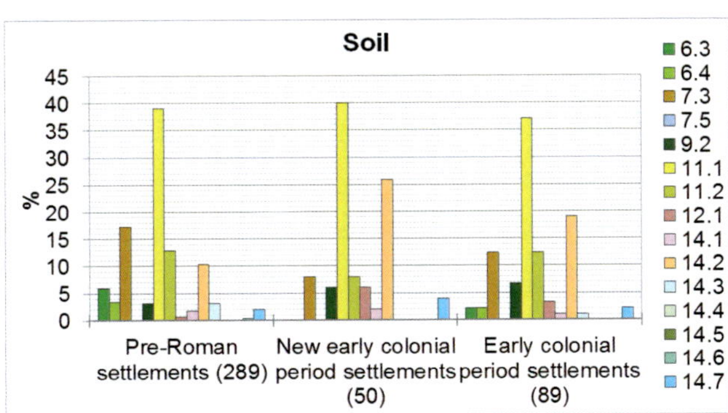

Soil unit	Area (sq m)	%
6.3	13156400	1.89
6.4	6690000	0.96
7.3	67255200	9.67
7.5	7361700	1.06
9.2	15808800	2.27
11.1	279720100	40.23
11.2	135817200	19.53
12.1	23297100	3.35

Soil unit	Area (sq m)	%
14.1	28993500	4.17
14.2	61494600	8.84
14.3	8919300	1.28
14.4	2542200	0.37
14.5	9745700	1.40
14.6	15408200	2.22
14.7	19072200	2.74
Tot.	695282200	100

Fig. B.4 Graphs of the percentages of pre-Roman, new Early colonial period, and Early colonial settlements with respect to altitude, slope, aspect and soil factors (cf. section 6.4.2 of Chapter 6). In brackets, number of total settlements per period.

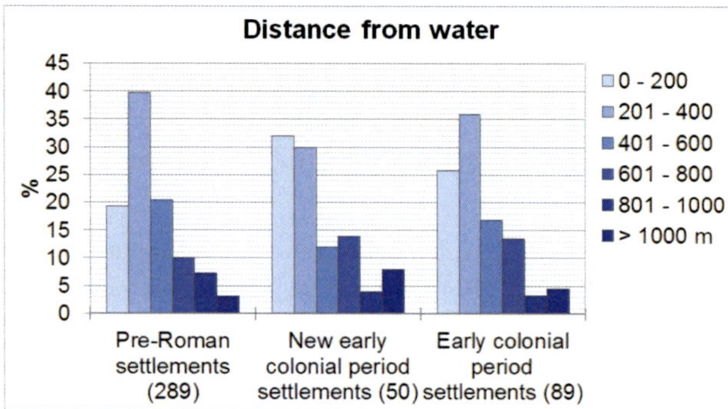

Water band		Area (sq m)	%
1st	0 - 200	175648200	25.26
2nd	201 - 400	164570200	23.67
3rd	401 - 600	127113900	18.28
4th	601 - 800	89448700	12.87
5th	801 - 1000	56637100	8.15
6th	> 1000 m	81864100	11.77
Tot.		695282200	100

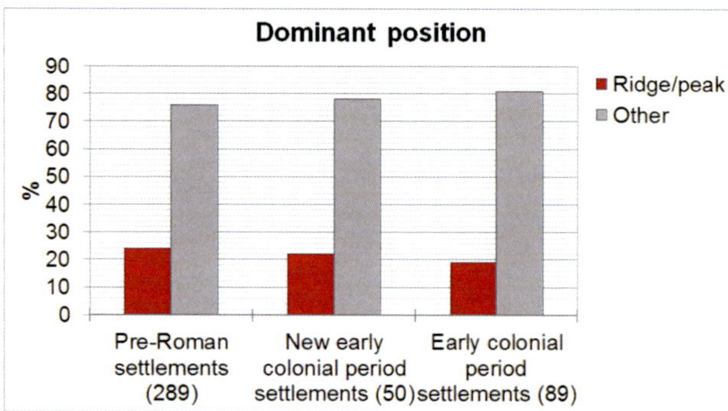

Dominant position	Area (sq m)	%
Ridge/Peak	129836500	18.67
Other	565445700	81.33
Tot.	695282200	100

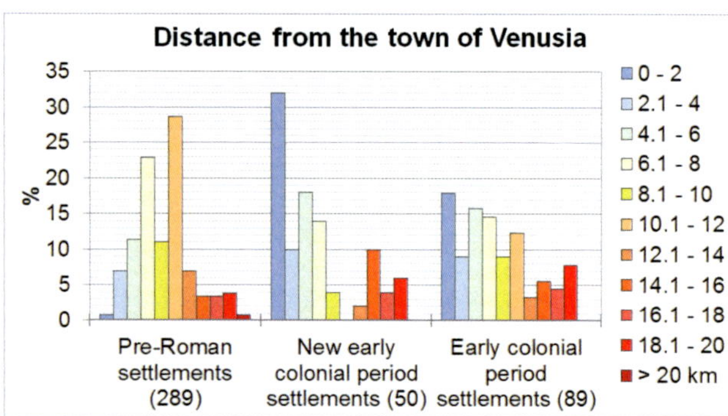

Distance band		Area (sq m)	%
1st	0 - 2	20050700	2.88
2nd	2.1 - 4	44602500	6.42
3rd	4.1 - 6	66629100	9.58
4th	6.1 - 8	82483500	11.86
5th	8.1 - 10	99370200	14.29
6th	10.1 - 12	97895100	14.08
7th	12.1 - 14	88587200	12.74
8th	14.1 - 16	60777500	8.74
9th	16.1 - 18	43986300	6.33
10th	18.1 - 20	41855200	6.02
11th	> 20 km	49044900	7.05
Tot.		695282200	100

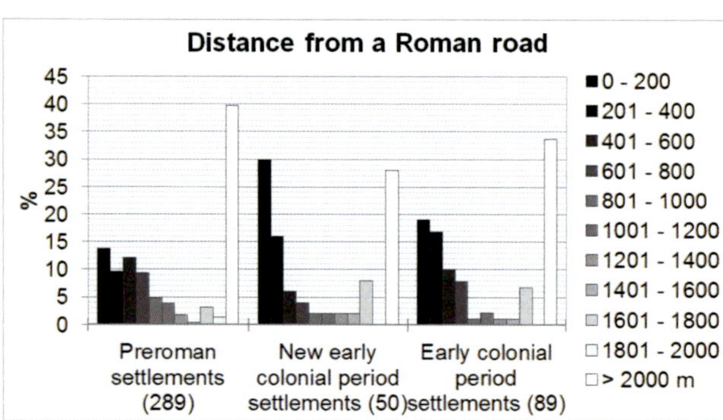

Distance band		Area (sq m)	%
1st	0 - 200	51136700	7.35
2nd	201 - 400	46892800	6.74
3rd	401 - 600	40638300	5.84
4th	601 - 800	37719400	5.43
5th	801 - 1000	34711500	4.99
6th	1001 - 1200	31034200	4.46
7th	1201 - 1400	27170200	3.91
8th	1401 - 1600	23640300	3.40
9th	1601 - 1800	21404100	3.08
10th	1801 - 2000	20094000	2.89
11th	> 2000 m	360840700	51.90
Tot.		695282200	100

Fig. B.5 Graphs of the percentages of pre-Roman, new Early colonial period, and Early colonial settlements with respect to water distance, the location of dominant positions, the distance from the town of Venusia and Roman roads (cf. section 6.4.2 of Chapter 6). In brackets, number of total settlements per period.

b) *Distribution graphs per period and per size*

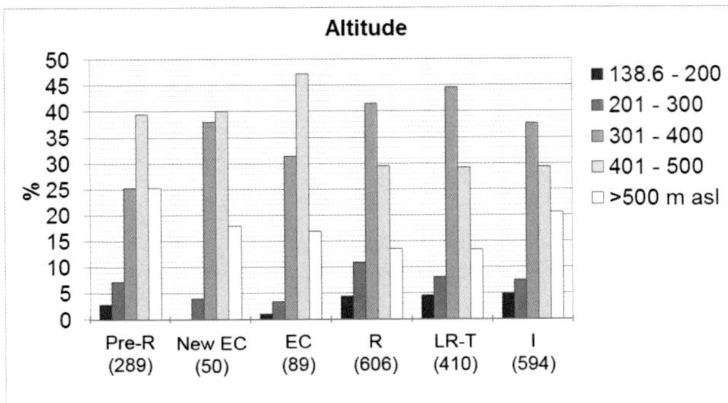

Altitude band		Area (sq m)	%
1st	138.6 - 200	42809500	6.2
2nd	201 - 300	78911100	11.3
3rd	301 - 400	236098000	34.0
4th	401 - 500	179970600	25.9
5th	> 500 m a.s.l.	157493000	22.7
	Tot.	695282200	100

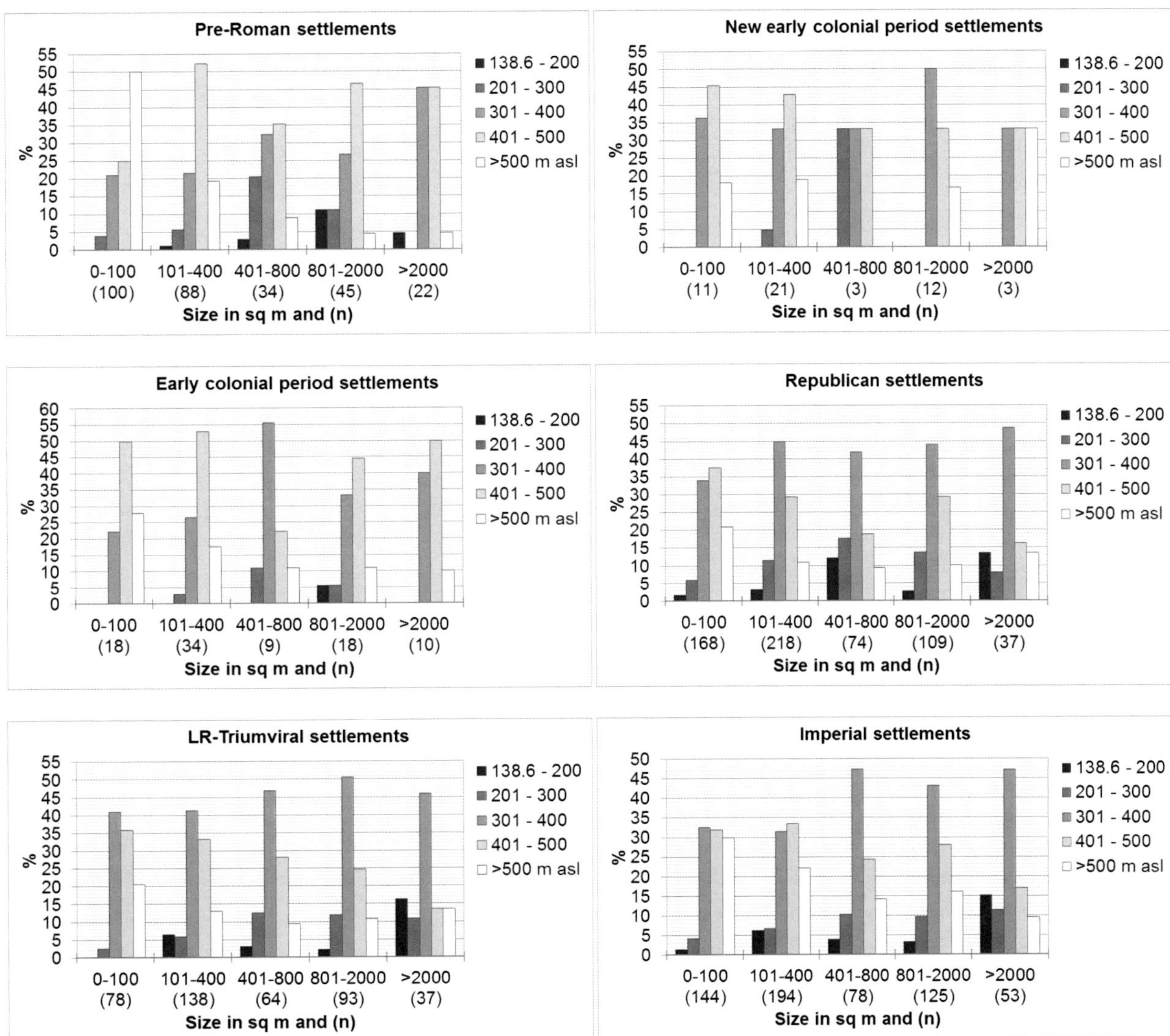

Fig. B.6 Settlement percentages with respect to altitude bands. In brackets, total number of settlements per sample (n). In the first top-left graph: Pre-R = Pre-Roman settlements; New EC = New early colonial period settlements; EC = early colonial period settlements; R = Republican settlements; LR-T = Late Republican-Triumviral settlements; I = Imperial settlements.

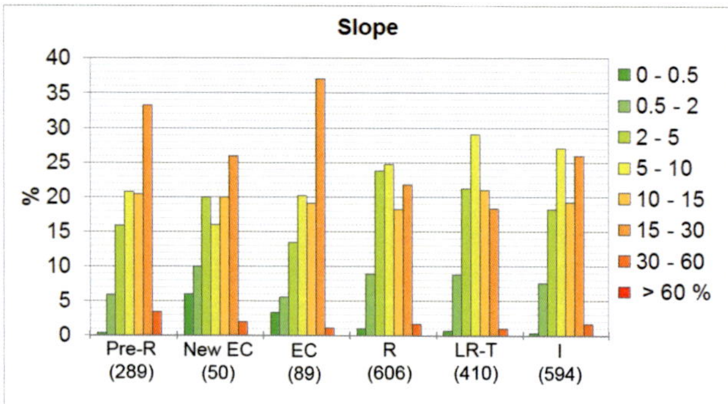

Slope class		Area (sq m)	%
1st	0 - 0.5	3679400	0.53
2nd	0.5 - 2	79234200	11.40
3rd	2 - 5	130012400	18.70
4th	5 - 10	121022700	17.41
5th	10 - 15	107931100	15.52
6th	15 - 30	201213300	28.94
7th	30 - 60	49488800	7.12
8th	> 60 %	2700300	0.39
Tot.		695282200	100

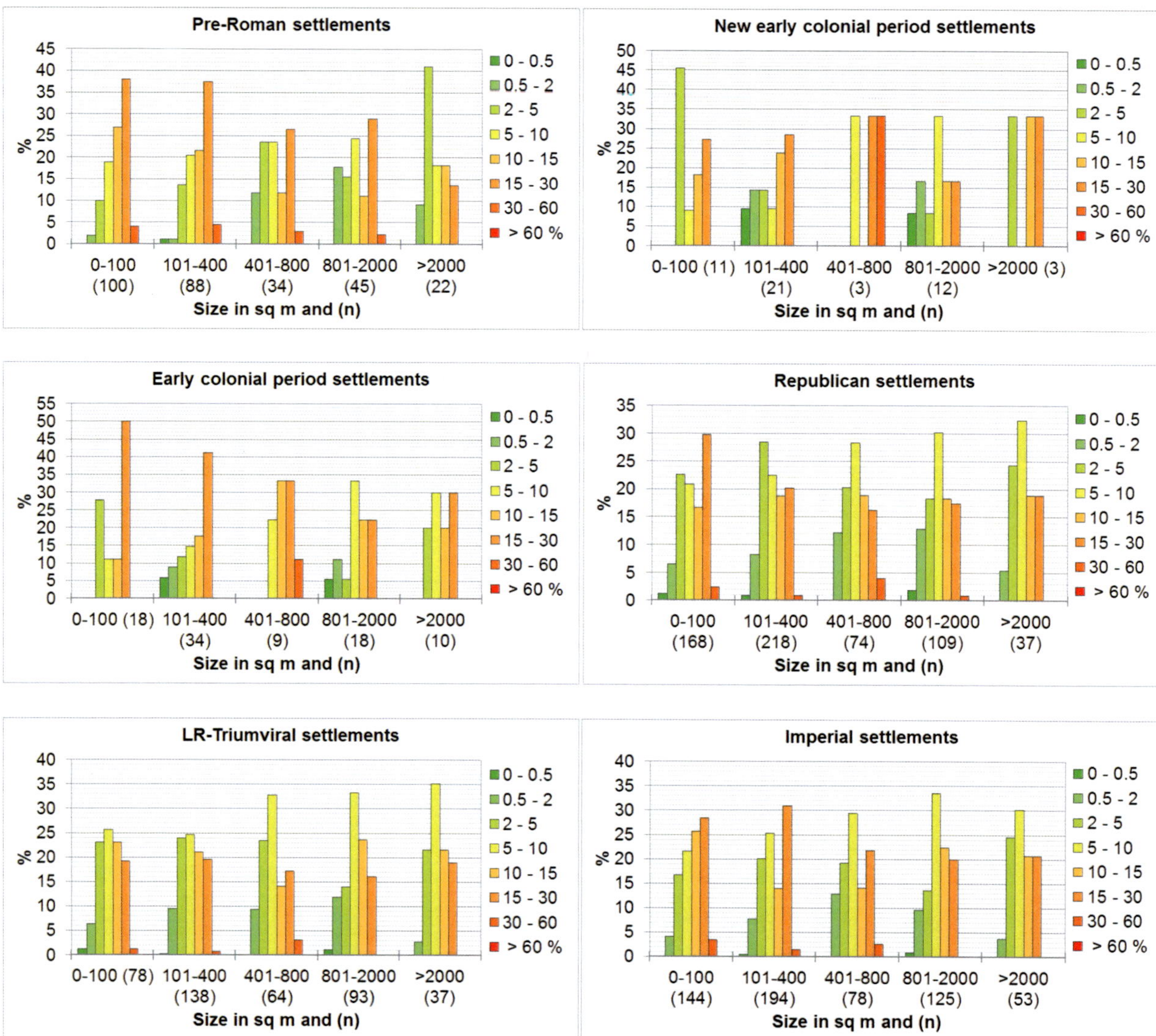

Fig. B.7 Settlement percentages with respect to slope classes. In brackets, total number of settlements per sample (n). In the first top-left graph: Pre-R = Pre-Roman settlements; New EC = New early colonial period settlements; EC = early colonial period settlements; R = Republican settlements; LR-T = Late Republican-Triumviral settlements; I = Imperial settlements.

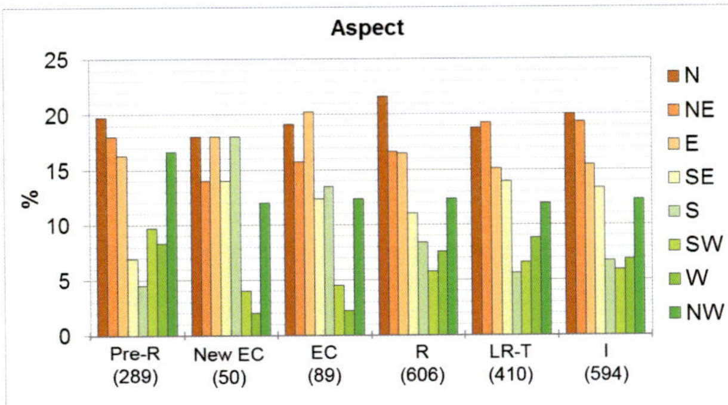

Aspect class		Area (sq m)	%
	Flat	63400	0.01
1st	N	143747900	20.67
2nd	NE	112804900	16.22
3rd	E	84013500	12.08
4th	SE	74520900	10.72
5th	S	60661100	8.72
6th	SW	55319500	7.96
7th	W	64981400	9.35
8th	NW	99169600	14.26
Tot.		695282200	100

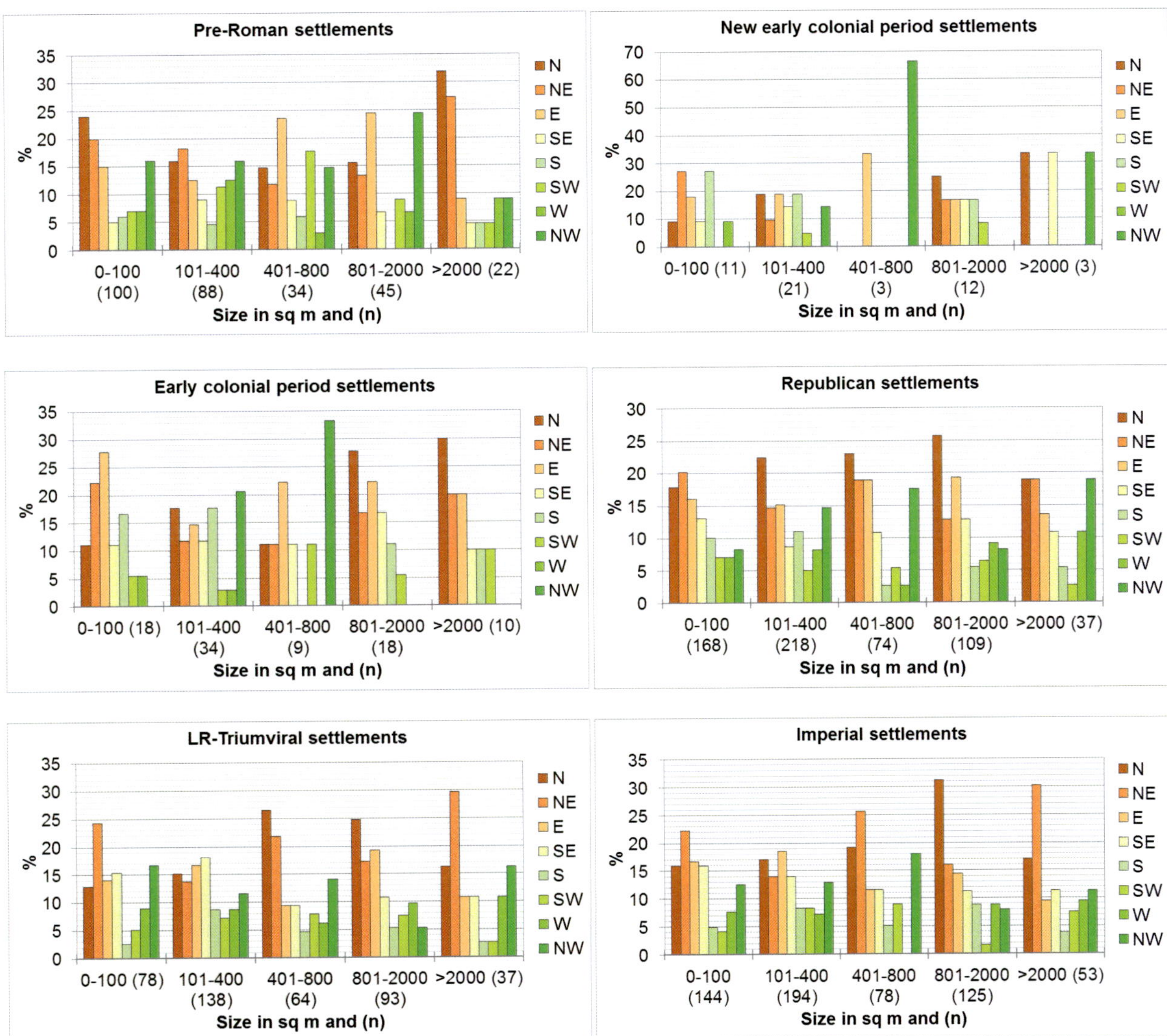

Fig. B.8 Settlement percentages with respect to aspect classes. In brackets, total number of settlements per sample (n). In the first top-left graph: Pre-R = Pre-Roman settlements; New EC = New early colonial period settlements; EC = early colonial period settlements; R = Republican settlements; LR-T = Late Republican-Triumviral settlements; I = Imperial settlements.

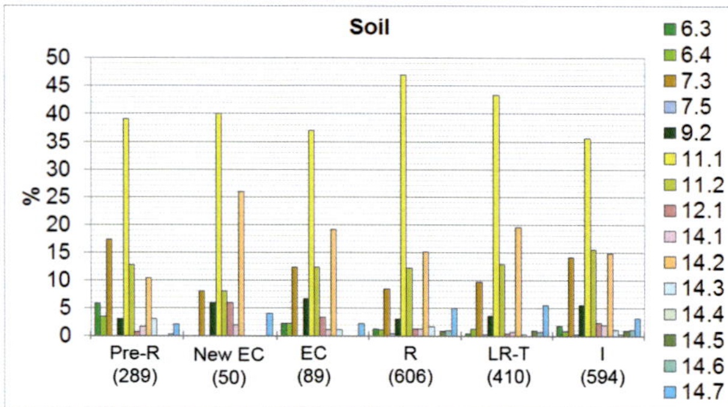

Soil unit	Area (sq m)	%
6.3	13156400	1.89
6.4	6690000	0.96
7.3	67255200	9.67
7.5	7361700	1.06
9.2	15808800	2.27
11.1	279720100	40.23
11.2	135817200	19.53
12.1	23297100	3.35

Soil unit	Area (sq m)	%
14.1	28993500	4.17
14.2	61494600	8.84
14.3	8919300	1.28
14.4	2542200	0.37
14.5	9745700	1.40
14.6	15408200	2.22
14.7	19072200	2.74
Tot.	695282200	100

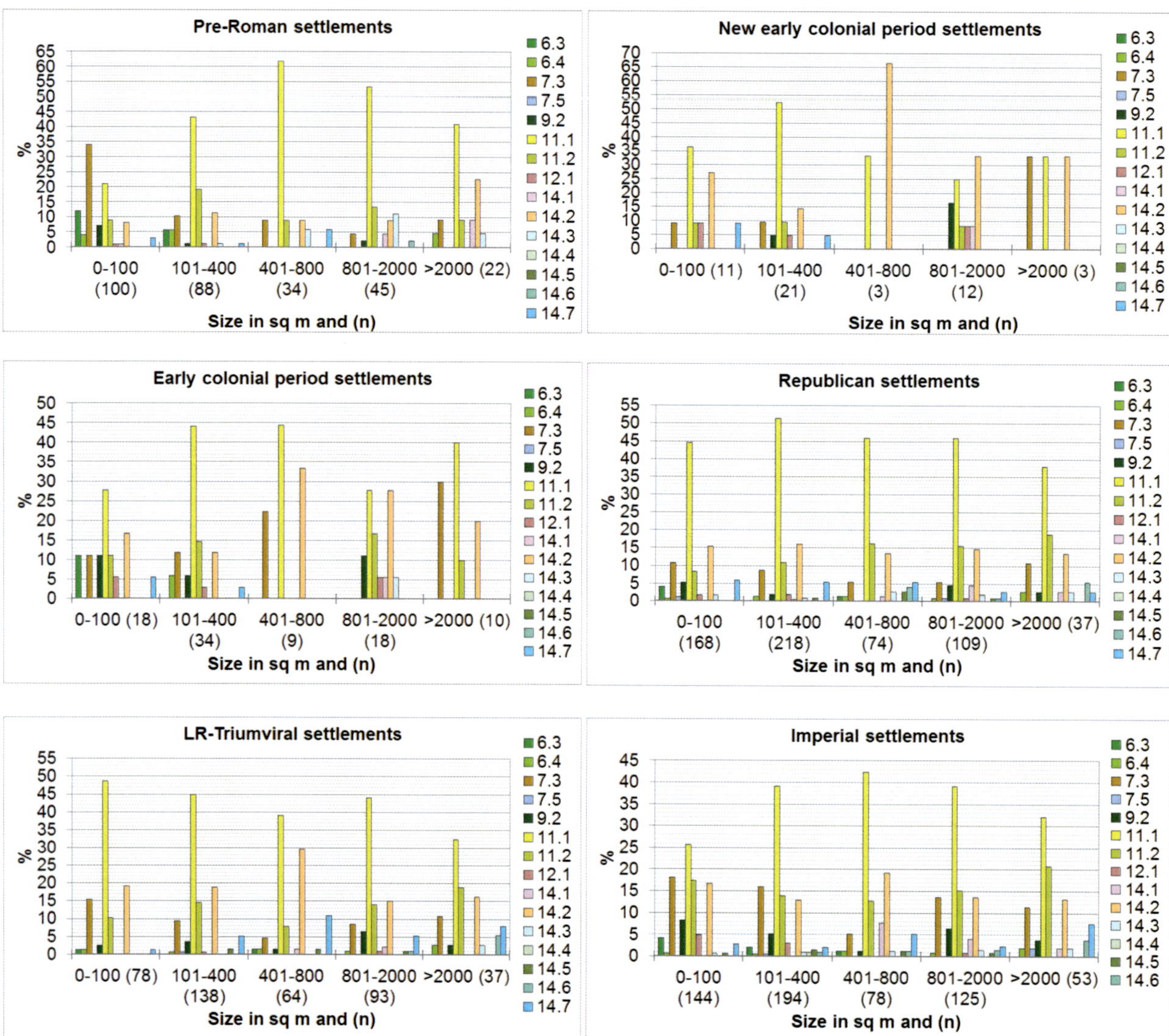

Fig. B.9 Settlement percentages with respect to soil units. In brackets, total number of settlements per sample (n). In the first top-left graph: Pre-R = Pre-Roman settlements; New EC = New early colonial period settlements; EC = early colonial period settlements; R = Republican settlements; LR-T = Late Republican-Triumviral settlements; I = Imperial settlements.

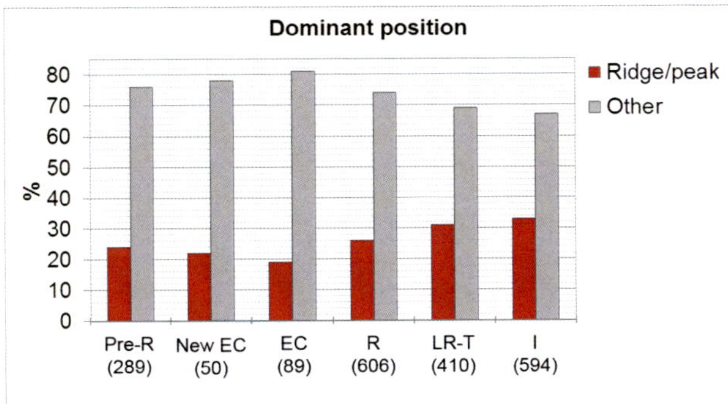

Dominant position	Area (sq m)	%
Ridge/Peak	129836500	18.67
Other	565445700	81.33
Tot.	695282200	100

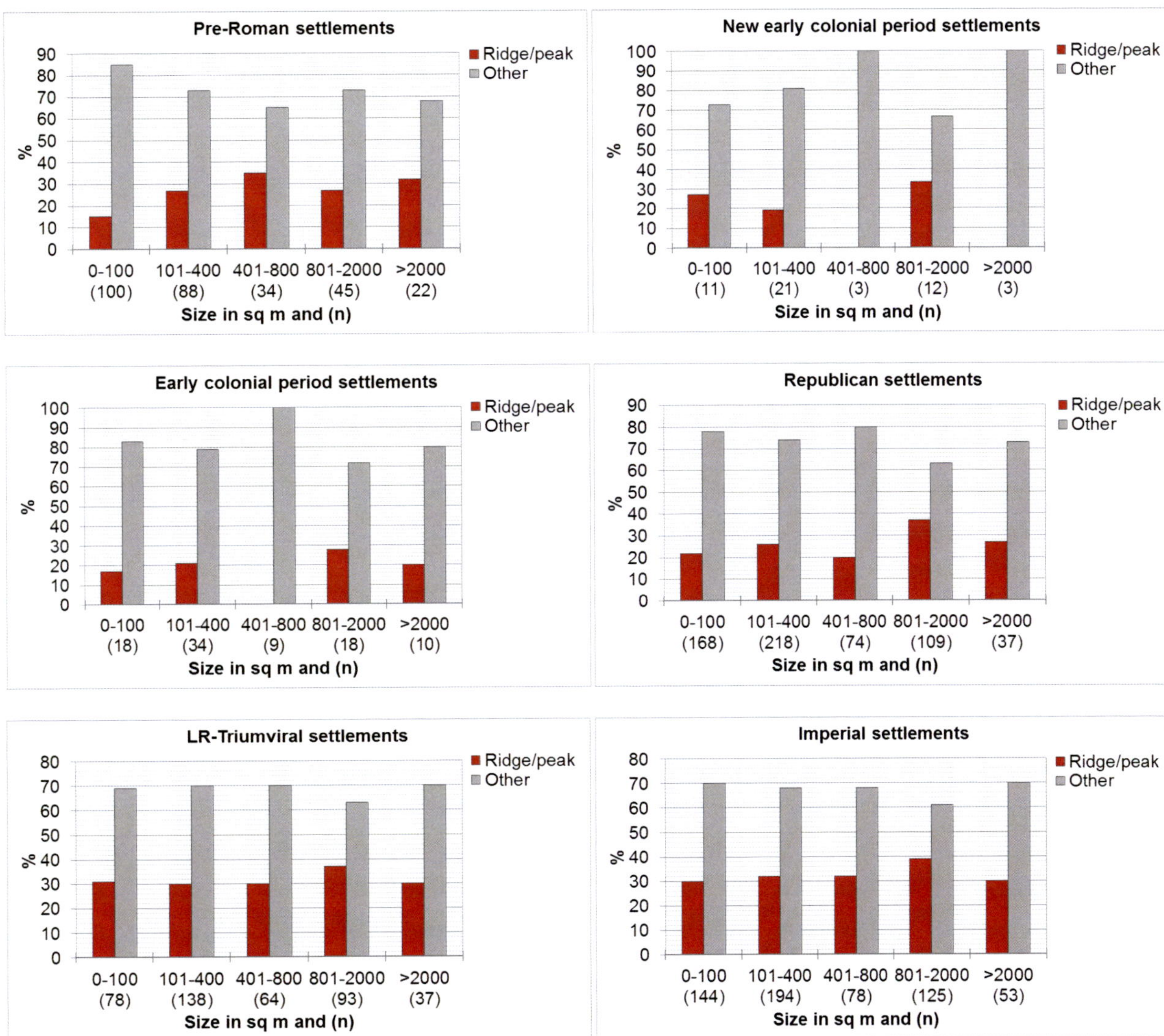

Fig. B.10 Settlement percentages with respect to dominant positions of the landscape. In brackets, total number of settlements per sample (n). In the first top-left graph: Pre-R = Pre-Roman settlements; New EC = New early colonial period settlements; EC = early colonial period settlements; R = Republican settlements; LR-T = Late Republican-Triumviral settlements; I = Imperial settlements.

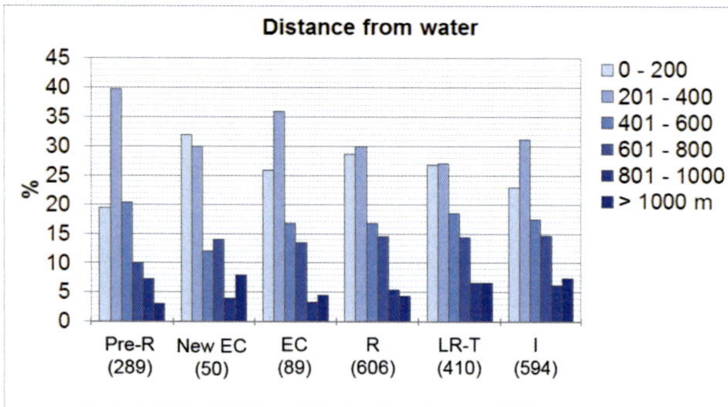

Water band		Area (sq m)	%
1st	0 - 200	175648200	25.26
2nd	201 - 400	164570200	23.67
3rd	401 - 600	127113900	18.28
4th	601 - 800	89448700	12.87
5th	801 - 1000	56637100	8.15
6th	> 1000 m	81864100	11.77
Tot.		695282200	100

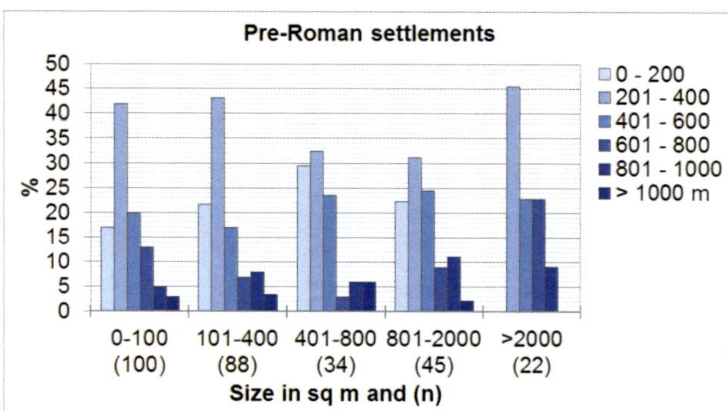

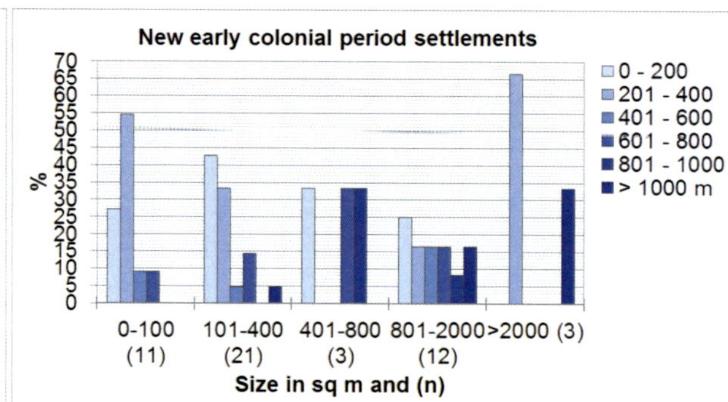

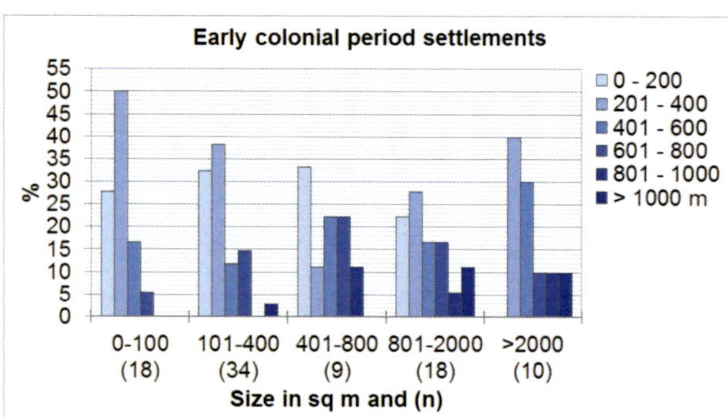

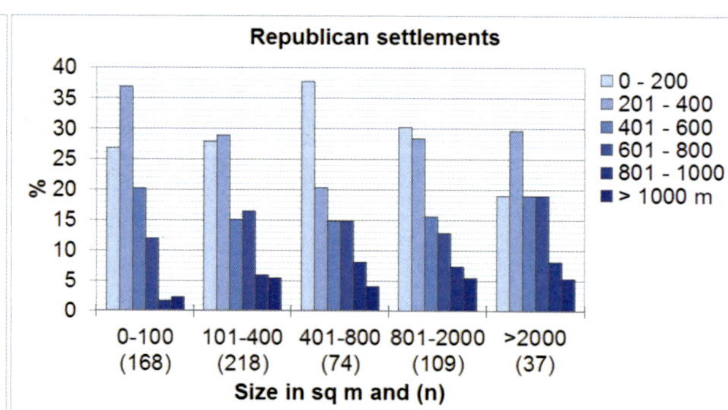

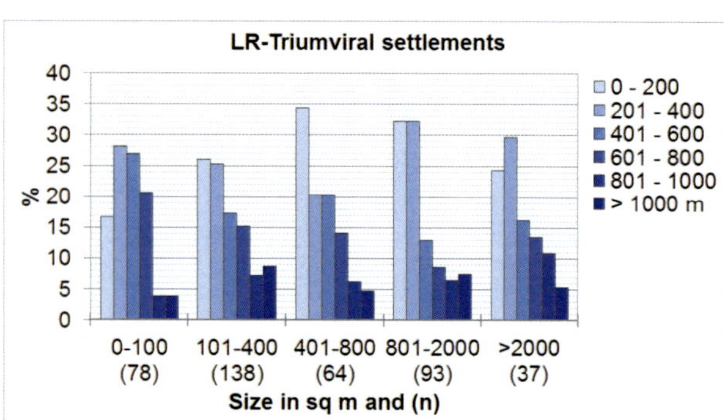

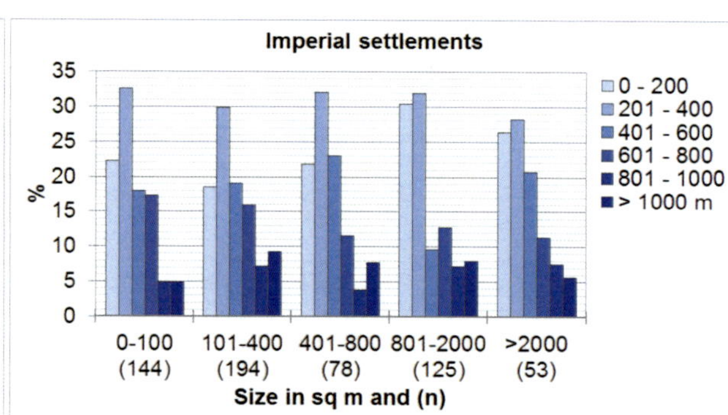

Fig. B.11 Settlement percentages in progressive distance bands from water sources. In brackets, total number of settlements per sample (n). In the first top-left graph: Pre-R = Pre-Roman settlements; New EC = New early colonial period settlements; EC = early colonial period settlements; R = Republican settlements; LR-T = Late Republican-Triumviral settlements; I = Imperial settlements.

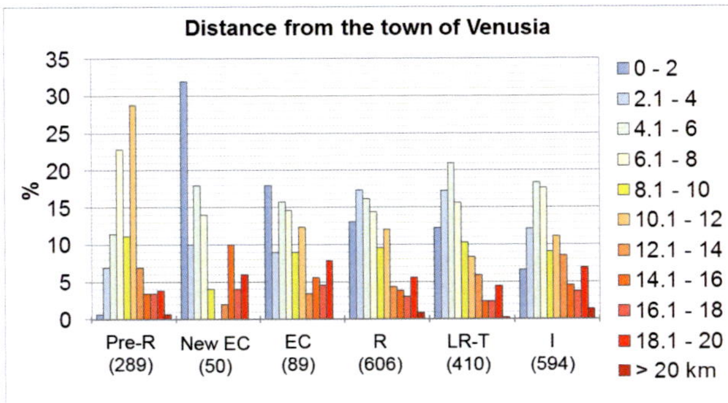

Distance band		Area (sq m)	%
1st	0 - 2	20050700	2.88
2nd	2.1 - 4	44602500	6.42
3rd	4.1 - 6	66629100	9.58
4th	6.1 - 8	82483500	11.86
5th	8.1 - 10	99370200	14.29
6th	10.1 - 12	97895100	14.08
7th	12.1 - 14	88587200	12.74
8th	14.1 - 16	60777500	8.74
9th	16.1 - 18	43986300	6.33
10th	18.1 - 20	41855200	6.02
11th	> 20 km	49044900	7.05
Tot.		695282200	100

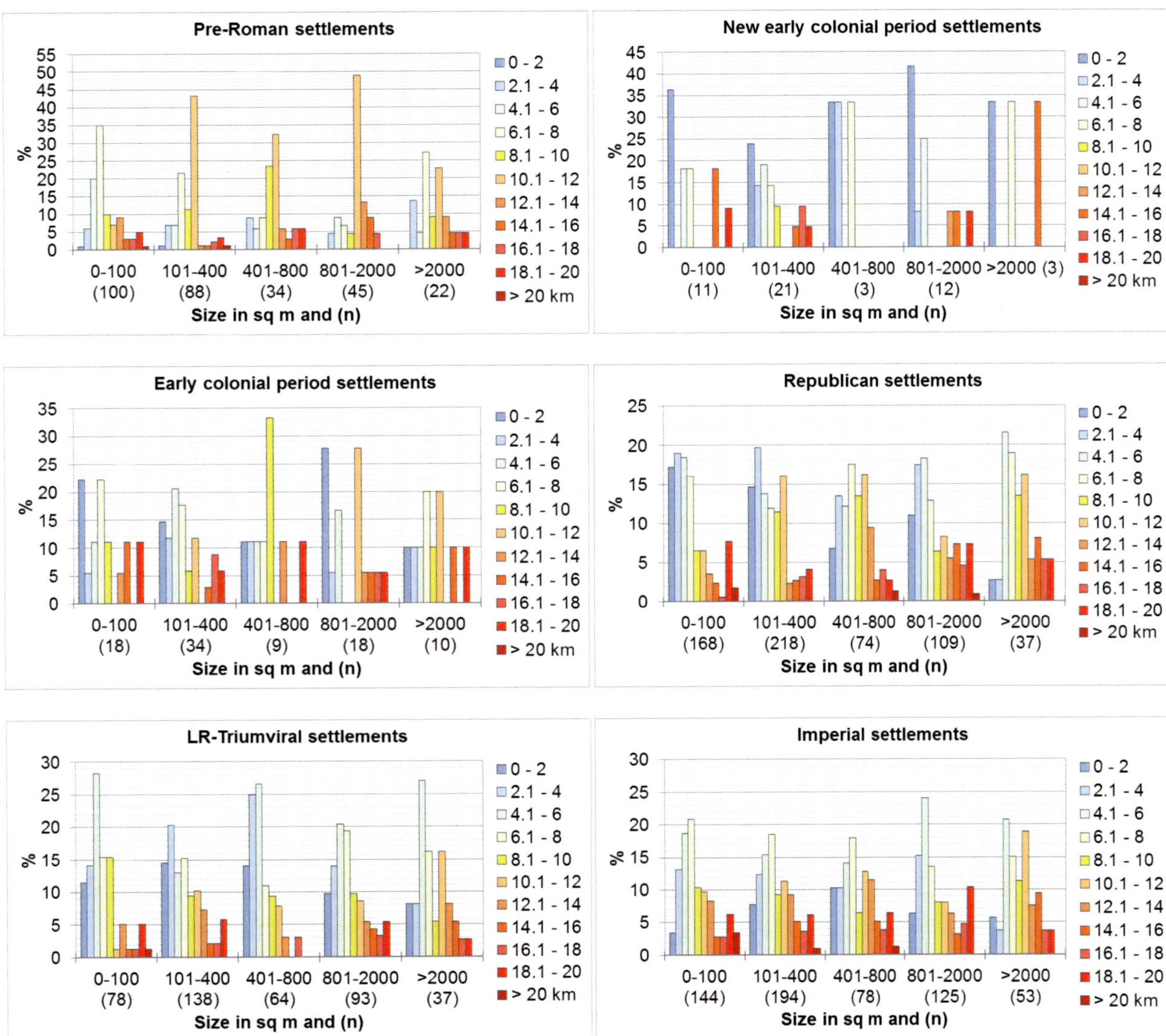

Fig. B.12 Settlement percentages in progressive distance bands from the town of Venusia. In brackets, total number of settlements per sample (n). In the first top-left graph: Pre-R = Pre-Roman settlements; New EC = New early colonial period settlements; EC = early colonial period settlements; R = Republican settlements; LR-T = Late Republican-Triumviral settlements; I = Imperial settlements.

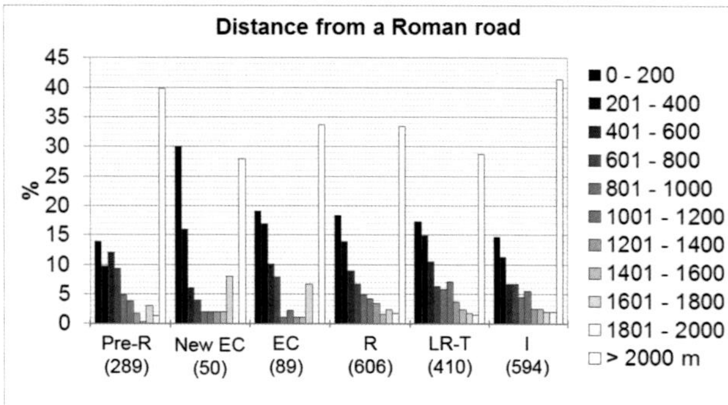

Distance band		Area (sq m)	%
1st	0 - 200	51136700	7.35
2nd	201 - 400	46892800	6.74
3rd	401 - 600	40638300	5.84
4th	601 - 800	37719400	5.43
5th	801 - 1000	34711500	4.99
6th	1001 - 1200	31034200	4.46
7th	1201 - 1400	27170200	3.91
8th	1401 - 1600	23640300	3.40
9th	1601 - 1800	21404100	3.08
10th	1801 - 2000	20094000	2.89
11th	> 2000 m	360840700	51.90
Tot.		695282200	100

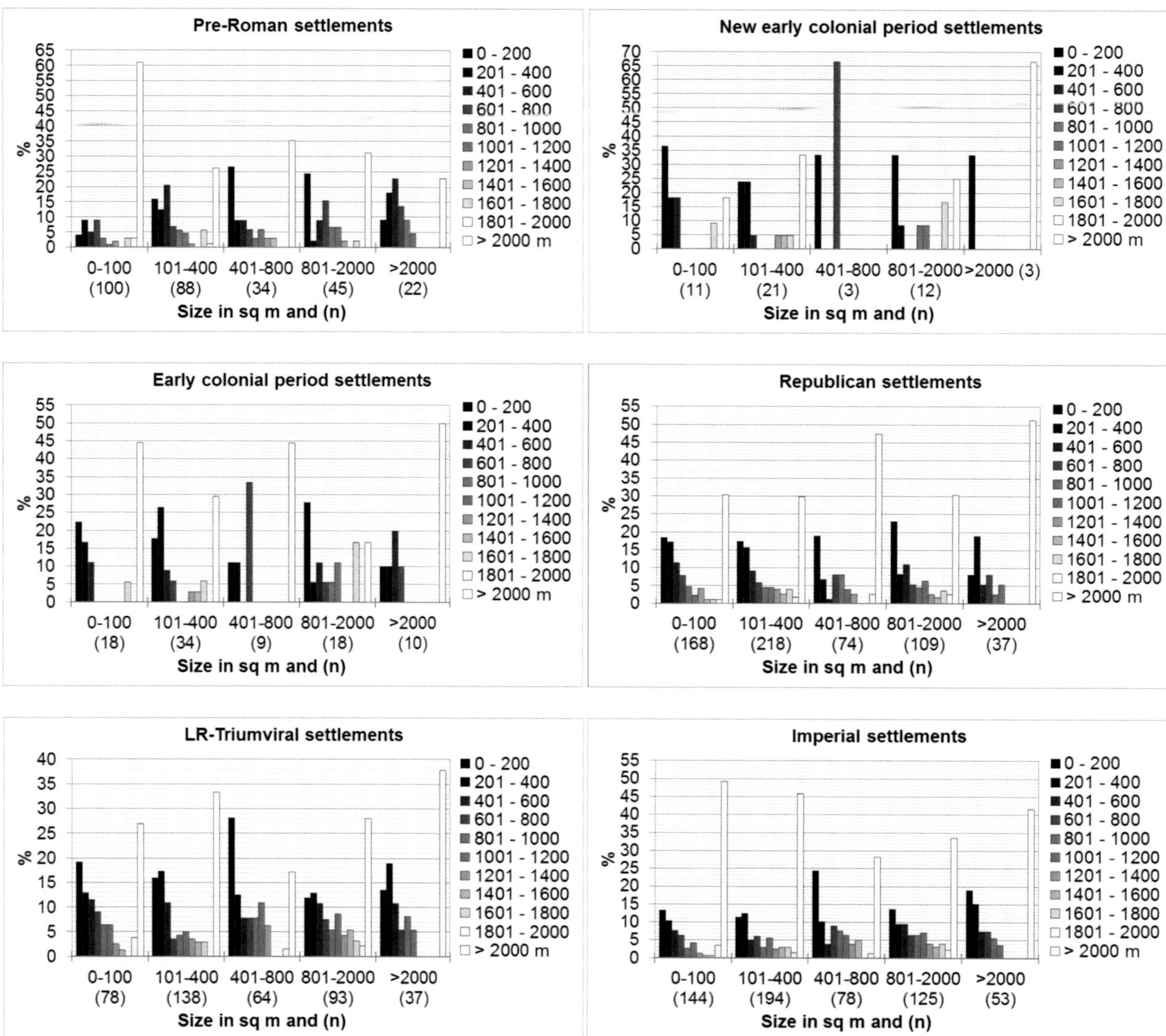

Fig. B.13 Settlement percentages in progressive distance bands from the (Roman) roads. In brackets, total number of settlements per sample (n). In the first top-left graph: Pre-R = Pre-Roman settlements; New EC = New early colonial period settlements; EC = early colonial period settlements; R = Republican settlements; LR-T = Late Republican-Triumviral settlements; I = Imperial settlements.

c) *Significant correlations identified by statistical tests*

The Chi-squared test, the Kolmogorov-Smirnov test and the Attwell-Fletcher test (for references see footnote 29, p. 113) were used to highlight possible significant correlations in the dataset. A significant level α of 0.05 was selected. The Chi-squared and the Attwell-Fletcher test were used in combination to identify significant correlations between settlement distributions and nominal variables (*e.g.* soils). Whereas, the Kolmogorov-Smirnov test and the Attwell-Fletcher test were used in combination to analyze settlement distributions with respect to ordinal variables (*e.g.* altitude). Chi-squared and Kolmogorov-Smirnov test allowed to assess whether a significant difference in frequency distribution exists between the observed and the expected number of sites. The Attwell-Fletcher test permitted to identify the land units where correlations likely occur[33]. In the tables below significant correlations are highlighted in black, absence of significant correlations is in white and the gray fields indicate that only one out of two tests detected a significant difference and/or correlation. The symbol "+" indicates that more sites than expected are located in a certain land unit (positive correlation, *i.e.* location preference), whereas "-" indicates fewer sites than expected (negative correlation, *i.e.* location disfavor). Positive and negative correlations are separated by a semicolon. The ordinal numbers indicate the classes/bands/units where the correlations occur (cf. Fig. 6.5, pp. 114-115).

33 For further methodological details on the application of these tests reference is made to Casarotto, A. 2017. Location preferences of rural settlements in the territory of Venusia: an inductive approach. In: H. Kamermans & C.C. Bakels (eds.), *Excerpta Archaeologica Leidensia II*, 163-209. Analecta Praehistorica Leidensia 47.

Table B.1 Location preferences of pre-Roman settlements.

Pre-Roman settlements	0-100 sq m	101-400 sq m	401-800 sq m	801-2000 sq m	>2000 sq m	Totality
Altitude	+ 5th ; - 1st	+ 4th ; - 1st				+ 4th ; - 1st
Slope						
Aspect						
Soil	+ unit 6.3	+ unit 6.4	+ unit 14.3	+ unit 14.3		+ units 6.3, 6.4
Ridges/Peaks						+ on ridges/peaks
Water	+ 2nd ; - 6th	+ 2nd ; - 6th				+ 2nd ; - 6th
Town	+ 3rd , 4th	+ 6th		+ 6th		+ 4th , 6th; - 1st , 11th
Roads		+ 3rd				+ 3rd ; - 8th

Table B.2 Location preferences of early colonial-period settlements.

Early colonial-period settlements	0-100 sq m	101-400 sq m	401-800 sq m	801-2000 sq m	>2000 sq m	Totality
Altitude	+ 4th	+ 4th				+ 4th ; - 1st
Slope		+ 1st				+ 1st
Aspect						- 7th
Soil						
Ridges/Peaks						
Water	+ 2nd	- 5th				+ 2nd
Town	+ 1st	+ 1st		+ 1st		+ 1st
Roads			+ 4th			

Table B.3 Location preferences of Republican settlements.

Republican settlements	0-100 sq m	101-400 sq m	401-800 sq m	801-2000 sq m	>2000 sq m	Totality
Altitude	+ 4th ; - 1st	+ 3rd				+ 3rd ; - 5th
Slope						
Aspect						
Soil						
Ridges/Peaks				+ on ridges/peaks		+ on ridges/peaks
Water	+ 2nd ; - 5th , 6th	- 6th				+ 2nd ; - 5th , 6th
Town	+ 1st , 2nd ; - 7th–9th, 11th	+ 1st , 2nd ; - 7th, 8th, 11th		+ 1st , 2nd		+ 1st , 2nd ; - 7th–9th, 11th
Roads	+ 1st , 2nd			+ 1st		+ 1st , 2nd ; - 8th

Table B.4 Location preferences of Late Republican-Triumviral settlements.

LR-Triumviral settlements	0-100 sq m	101-400 sq m	401-800 sq m	801-2000 sq m	>2000 sq m	Totality
Altitude	+ 4th ; - 1st			+ 3rd	+ 1st	+ 3rd
Slope						
Aspect						
Soil						+ unit 14.2
Ridges/Peaks		+ on ridges/peaks		+ on ridges/peaks		+ on ridges/peaks
Water						- 6th
Town	+ 1st, 3rd	+ 1st, 2nd	+ 1st, 2nd	+ 1st		+ 1st–3rd ; - 6th–9th, 11th
Roads		+ 2nd				+ 1st, 2nd

Table B.5 Location preferences of Imperial settlements.

Imperial settlements	0-100 sq m	101-400 sq m	401-800 sq m	801-2000 sq m	>2000 sq m	Totality
Altitude	+ 5th ; - 1st				+ 1st	
Slope						
Aspect						
Soil	+ unit 9.2					+ unit 9.2
Ridges/Peaks	+ on ridges/peaks	+ on ridges/peaks	+ on ridges/peaks	+ on ridges/peaks		+ on ridges/peaks
Water	- 6th					+ 2nd ; - 6th
Town	+ 2nd	+ 1st ; - 11th	+ 1st	+ 3rd		+ 1st–3rd ; - 8th, 11th
Roads						+ 1st

FSC
www.fsc.org
MIX
Papier | Fördert gute Waldnutzung
FSC® C083411

Kolibri 360 GmbH
Ferdinand-Jühlke-Straße 7
99095 Erfurt, Deutschland
produktsicherheit@kolibri360.de